Water and Ecosystems:

Managing Water in Diverse Ecosystems to Ensure Human Well-being

Water and Ecosystems

Managing Water in Diverse Ecosystems to Ensure Human Well-being

Edited by
Caroline King
with
Jennifer Ramkissoon,
Miguel Clüsener-Godt
and Zafar Adeel

UNESCO
UNITED NATIONS UNIVERSITY
UNU-INWEH

Disclaimer

The designations employed and presentations of material throughout this publication do not imply the expression of any opinion whatsoever on the part of the United Nations Educational, Scientific and Cultural Organization (UNESCO) and the United Nations University (UNU) concerning legal status of any country, territory, city or area or of its authorities, or concerning the delimitation of its frontiers or boundaries. The views expressed in this publication are those of the respective authors and do not necessarily reflect the views of the UNESCO or UNU. Mention of the names of firms or commercial products does not imply endorsement by UNESCO or UNU.

Available from:

United Nations University
International Network on Water,
Environment and Health (UNU-INWEH)
50 Main Street East, First Floor,
Hamilton, Ontario
Canada L8N 1E9
Tel: +1-905-525-9140, ext. 24517
Fax: +1-905-529-4261
Email: contact@inweh.unu.edu

Cover photos: Caroline King and Guan Zhuojin

ISBN 92-808-6002-X

Contents

Foreword

This compilation of papers is derived from an International Workshop co-hosted by the United Nations University's International Network on Water, Environment and Health (UNU-INWEH) and UNESCO-MAB-IHP in June, 2005. The topic addressed by the workshop was: "Water and Ecosystems: Water Resources Management in Diverse Ecosystems and Providing for Human Needs". UNU-INWEH was privileged to host the workshop at McMaster University, Hamilton, Canada. We welcomed case study presenters from the Amazon, the Arabian Gulf, East Africa, India, China, Europe and Canada, as well as thematic presenters and discussants from a number of other countries.

The topic of the Workshop and of these Proceedings is important, in my view, for at least two reasons.

First, the Workshop's theme lies at an important boundary - an often awkward boundary - between the two dominant challenges that we face in addressing the global water crisis. Many will recall the disjunct that developed at the World Summit on Sustainable Development at Johannesburg. To oversimplify somewhat, on the one side of the boundary was the "safe water-health" camp, who wanted a preemptive priority placed on safe drinking water and sanitation. Given the horrific scale and brutality of the problem, it was hard not to sympathize with them. More than 2 1/2 billion people lack access to sanitation and nearly half that many lack safe drinking water. As a result, over 3 million people, mostly children, die each year from water-related health problems, while billions are made ill from dysentery.

On the other side of the boundary was the "ecological integration" camp, arguing that isolated action was fatal, that water supply and sanitation must be considered within the broader framework of IWRM, not only to succeed in providing safe drinking water and sanitation, but also to protect aquatic ecosystems and ecosystem services. How else, they pleaded, can all of the other important water challenges be met,. such as water pollution, basin degradation, urban mega-city issues, dams and diversions, marine coastal pollution, transboundary water challenges, groundwater pollution and climate impacts?

In the end, the Johannesburg Summit committed the world to the two MDGs for safe drinking water and sanitation by 2015, and, on the ecosystem integration side, committed all countries to produce IWRM plans by 2005. Now that we are beyond 2005, where do we stand? While some modest progress has been made on the MDGs, there is almost total unanimity that we will not even come close to meeting the 2015 targets. And what of the progress on IWRM? I think it is correct to say that not a single country on earth produced their national plan by the end of 2005. At best, a number of countries started their planning in 2005!

Given this reality, it seems fair to conclude that we are failing on both sides of the boundary. Does this not suggest that the boundary itself may be at least a good part of the problem? If so, then the challenge we set for the workshop was valid: how can we reconcile these two solitudes?

My second brief, and somewhat simplistic, point has to do with human capacity and the water crisis. It is a truism that the water crisis is global in scale only in the aggregate. Water problems are inherently local in nature, addressed by local practitioners who are rooted in their local environment and collaborating locally on the complex web of water issues they face.

What this means, I think, is that the "what" cannot be divorced from the "who". The human expertise we need cannot be treated as a simple downstream resource, a routine tool, addressed after the plan is developed. In the developing world, there is a very intimate, interactive and iterative relationship between the "practitioners" and the "practice". In a fundamental way, people are the plan. If this is true, then a stronger focus on capacity development issues may offer new insights, new models, or new generalizations, on how to move forward. The workshop was challenged to consider this possibility during its deliberations.

I recommend these Proceedings to all who are committed to truly integrated water resources management.

Thank you.

Dr. Ralph J. Daley
Former Director, UNU-INWEH
(October 1996 – June 2006)

Preface

The International Workshop from which this book developed, "Water and Ecosystems: Water Resources Management in Diverse Ecosystems and Providing for Human Needs", was jointly organised by the United Nations University's International Network on Water, Environment and Health (UNU-INWEH) and UNESCO's Man and the Biosphere Programme (MAB) and International Hydrological Programme (IHP). This important meeting brought together many important partners in the protection and management of water resources in selected ecosystems. It was our pleasure to see that UNU, both at Headquarters as well at its offices worldwide, and its partners, such as McMaster University, are constantly co-operating with UNESCO. This ongoing cooperation includes a range of activities on different programmes for the conservation and sustainable development of coastal areas, as well as cooperation on issues relating to water and ecosystems, which is the principal priority in the UNESCO Science Sector.

This book follows up on an earlier book on 'Issues of Local and Globe Use of Water From the Amazon', Aragon and Clüsener-Godt, Eds., published by UNESCO in 2004. The previous year, 2003, was designated by the UN to be the International Year of Water, considering the strategic character of this resource for the new century. A series of events subsequently launched around the world have brought full attention to that theme. The figures are really alarming: 97.5% of the existing water on Earth is salty water, and only part of the existing fresh water (2.5% of the total) is useable. Only 1% of the fresh water on Earth is easily accessible, and the Amazon holds around 15% of this percentage. Water is also one of the most unequally distributed natural resources in the world: more than 40% of river waters, reservoirs and lakes are concentrated in six countries: Brazil, Russia, Canada, United States, China and India.

Therefore, it seemed urgent to try to evaluate the water resources management in diverse ecosystems and its provisions for human needs, which was the topic selected for our workshop, and for this book. Figures given for human consumption of water raise concern. In North-America and Europe, for example, citizens have around 140 litres/day of water for their own use, in developing countries of Africa only 10 litres/day of water are available per capita. A very scary estimate is given worldwide, saying that 1.5 billion people on this planet have already no access to safe potable water within easy reach of their homes. This means that, if urgent measures are not taken, two thirds of humanity will encounter water shortages by the year 2025.

It is up to us, the scientific community, to try to participate actively in the process of elaborating measures to solve the problem, which is becoming one of the most relevant geopolitical issues of our century. I would like to refer in this context to the

South-South Co-operation Programme for Ecodevelopment as an initiative of the MAB/UNESCO Programme, the United Nations University and the Third World Academy of Sciences, which operates from the Division of Ecological and Earth Sciences of UNESCO in Paris. This Programme with our partners is enabling the constant intellectual work of countries from the South to develop conceptual frameworks.

In this context, I am pleased to announce a UNESCO Chair for South Co-operation on Sustainable Development has been created recently at the University of Para, Belem, Brazil. This Chair will handle issues related to South-South Co-operation on tropical forests in the Amazon and the Congo Basin, as well as in the forests of South-East Asia.

In this book, we are starting to develop a framework for integrated water management in ecosystems and its provisions for human needs. This also leads us into a discussion on poverty reduction and ecological sustainability in water management. I do sincerely believe that the conclusions and recommendations presented in this book will bring us to some substantive follow-up activities.

The importance of international co-operation for the production and dissemination of knowledge related to the sustainable use of water resources is crucial. This was done for the Amazon region in an earlier publication by UNESCO (Aragon and Clüsener-Godt, 2004) and subsequently also for the Arab Gulf region, with a second joint UNESCO and UNU-INWEH publication (Amer et al., 2006). Nevertheless, continued South-South Co-operation, including universities, research institutes and NGOs in the production and dissemination of knowledge related to water resources is needed. The many examples of inter-university co-operation already existing in the Amazon, and in other regions of the world, are of utmost importance.

The necessity of jointly organizing these meetings and workshops is vital for the success of programmes dealing with water issues. Each agency, NGO, or Governmental Institution has its own strength. UNESCO's role is to try to merge, as much as possible, these converging forces into one strategic action plan and to establish a critical mass to tackle these important problems that humanity is increasingly faced with. So far, these initiatives have worked out well. The international workshop on "Water and Ecosystems: Water Resources Management in Diverse Ecosystems and Providing for Human Needs" was a living example of how real-life co-operation can exist and give concrete results at all levels of society, including the scientific community, the protected area managers, political-decision makers at all levels and stake-holders living in the areas concerned.

In this context, let us hope that we will also be able in the future to implement challenging programmes and projects that many developing countries are soliciting from us, the UN System. Whether we handle problems of management of water resources, or whether we observe still existing national boundary problems between neighbouring countries, the UN System and its partners have many concepts and approaches to provide in order to help to overcome these problems. Transboundary biosphere reserves, international scientific training and capacity building programmes are just a few examples. As we have learned from our experiences throughout the world, there are many ways that international scientific programmes work to support efforts to improve the world in which we live. UNESCO, through its main programmes such as MAB, the International Hydrological programme (IHP), or the natural part of the World Heritage Convention, is constantly engaged in such work. These efforts are achieved by UNESCO working side by side with its Member States in order to mobilize the international scientific community or to establish new large extra-budgetary projects financed mainly by donor countries, as it is the case with the World Water Assessment programme, which is financed by Japan.

I would like here to reiterate UNESCO's commitment to work with all its strength to try to alleviate the water-related problems that the world is facing at the beginning of this century. The scientists from all over the world who have contributed to this book are working to improve our understanding and conservation of our water and ecosystems. This will help us to ensure better living conditions, particularly of the poorest, who often lack basic access to appropriate water resources and, therefore, suffer from water-born diseases. I am confident that we all will continue on the same fruitful basis for co-operation, and, together, we shall achieve progress towards the goal of sustainable development.

Dr. Miguel Clusener Godt
Division of Ecological and Earth Sciences, Man and the Biosphere Programme (MAB), UNESCO

References

Aragón, L.E. and M. Clüsener-Godt (eds). (2004) *Issues of local and global use of water from the Amazon*, Montevideo: UNESCO, NAEA/UFPA.

Amer, K. M., B. Boer, M.C. Brook, M. Clüsener-Godt, and W. Saleh (eds). (2006) *PolicyPerspectives for Ecosystem and Water Management in the Arabian Peninsula*: UNESCO, UNU-INWEH.

Acknowledgements

The papers in this book are the result of an international workshop organized by the United Nations University's International Network on Water, Environment and Health (UNU-INWEH), and the United Nations Educational, Scientific and Cultural Organization (UNESCO) through its Man and Biosphere Programme (MAB) and International Hydrological Programme (IHP). Many international experts took part in this workshop. They are listed at the end of this volume. Their lively participation and informed reflections were invaluable to the development of the case studies and conclusions that are presented in this book.

The success of the international workshop was enabled by the coordinated efforts of the staff of UNU-INWEH and UNESCO. Many thanks are due to Ms Maud de Jorna, UNESCO and Ms Ann Caswell, Ms Corenne Hammond Agboraw and Ms Annette Dubreuil, UNU-INWEH. Ms Anne Ollivier, Intern, UNU-INWEH, played a special role in ensuring the success of the meeting through her care and attentions to all participants and colleagues before, after and during the workshop in Hamilton. Ms Leslie Kulperger, volunteer, also gave valuable assistance during the meeting.

Ms. Jennifer Ramkissoon, Intern, UNU-INWEH, has undertaken the tasks of formatting and reviewing the copy-editing of this volume with great patience and application. She has also provided many insightful suggestions to improve the final presentation of the book and its content.

Introduction

Conserving Ecosystems to Meet the Human Water Demands

Zafar Adeel and Caroline King, UNU-INWEH

Lack of safe water to serve the household needs in developing countries, and the related lack of access to sanitation facilities, has emerged as a major global crisis. This is amply demonstrated in the recently released Human Development Report (UNDP, 2006), which cites figures of over 1 billion people without access to clean water, over 2.6 billion without access to adequate sanitation, and 1.8 million children killed each year by preventable water-borne diseases. It is also quite significant that this report, which typically focuses on economic and social development issues, has singled out the water crisis. This truly reflects the growing international recognition of the importance of this crisis. Most commonly, this global crisis is described in terms of the lack of social and economic growth in developing countries. However, this crisis can also be related to the state of the ecosystems that are the source of the water needed to address the crisis. Not by coincidence, these ecosystems impose a water demand of their own, and provision of other benefits from various ecosystems to the human society is dependent on adequate water availability. Set against this backdrop of ecosystem conservation to meet the global water crisis, this book considers ways to meet the challenge of providing increased access to water supplies for basic human needs and socio-economic development. This is important because projections of future water demand and availability have already highlighted that greater stresses are likely to be placed on freshwater supplies as all regions of the world continue to develop.

This global crisis can also be viewed in the context of efforts to meet the Millennium Development Goals. In 2005, two major global reports highlighted the essential interrelatedness of the achievement of the two Millennium Development targets for sustainable access to safe drinking water and sanitation. The first was the report of the Millennium Project Task Force on Water and Sanitation (2005), which examined the prospects for achievement of the water and sanitation target, and the second was the Millennium Ecosystem Assessment (MA), which launched a global assessment of freshwater systems (Vorosmarty et al., 2005). Both of these reports observed that over-extraction of fresh water has profound long-term consequences for natural systems' ability to continue to supply essential services such as food, vegetation and waste assimilation, as well as the global water and climate cycles. Both reports concur that in order to safeguard and maintain the supply of fresh water for human needs, the management of fresh water supplies and the management of ecosystems must be effectively integrated. But what will this mean to water managers at the local and regional levels who are confronted with the challenges of water and ecosystem management? A diverse array of ecosystem conditions and population dynamics will shape the problems, possibilities and potential solutions available.

This book reflects on the challenge, viewed from different perspectives and different parts of the world.

The contributions included in this book primarily comprise papers presented at a workshop organized in Hamilton, Ontario (14-16 June, 2005) jointly by the United Nations University's International Network on Water, Environment and Health (UNU-INWEH) and UNESCO's Man and Biosphere Programme (MAB) and International Hydrological Programme (IHP). The workshop was entitled 'Water and Ecosystems: Water Resources Management in Diverse Ecosystems and Providing for Human Needs'. This workshop brought together an international group of scientists and researchers to discuss the need for integrated management of water resources and ecosystems. A series of case studies was developed for this purpose in order to juxtapose experiences from around the world in the pursuit of integration between water and ecosystem management. The workshop participants sought to identify transferable generic lessons from the case studies, as well as to highlight contrasts between differing regional contexts. In particular, the workshop participants considered the current management regimes in the various regions, and the present state of integration between management of water resources and ecosystems. By identifying gaps in the integration of these domains, they offered a series of recommendations concerning the needs for human and institutional capacity development.

Effective approaches to water and ecosystem management can be harnessed through ecologically-based methods that combine scientific understanding of ecosystem processes with an appreciation of the needs and involvement of local communities. Such approaches may be small in scale, led and catalyzed by participatory scientific research initiatives. The successful implementation of improvements in water and ecosystem management of this nature depends on the availability of human capacity. This includes both the local availability of sufficient human capacity to develop and implement management solutions on the one hand, as well as a broader level of institutional and political support, on the other. Previous publications developed by UNU-INWEH point to the lack of capacity – human, institutional, scientific and technological – in developing countries as a key difficulty in achieving such integration (Daley and Dowdeswell, 2002). It was demonstrated in these earlier studies that the availability of management capacity was indeed inversely related to the prevalence of environmental problems in the areas under study. These findings coincide with studies of water management capacity conducted at the local level in the US and Canada (Pirie et al., 2004, Ivey et al., 2004). UNU-INWEH has accordingly developed a generic conceptual approach to defining capacity development needs for water resources management according to these four categories, or pillars (Daley and Dowdeswell, 2002).

Thematic focus of the book

The first section of this book offers a conceptual discussion on approaches to water resource management in diverse ecosystems, drawing on the findings of the MA and other ecosystem-oriented initiatives. The first two chapters present global reviews of human water needs and available freshwater resources. Both chapters draw on the findings of the MA, and pursue issues raised by this recent assessment. The first paper, by Revenga, considers global conditions and trends in freshwater ecosystems and the challenges to meet human water needs through water resources development. Revenga reflects upon the existing stresses on freshwater ecosystems, and the exacerbation of stresses that is likely to result from the increasing demands. In order for managers to understand and mitigate these stresses, and to optimize the benefits of management decisions, Revenga emphasizes the need to collect better information and data on freshwater ecosystems.

The second chapter, by Finlayson, focuses in particular on human needs for water for agriculture. It examines the imperative to balance these water needs for human use against the needs in wetland ecosystems to retain water for nature protection. In seeking this balance between two almost conflicting needs for water, Finlayson also considers the scientific challenges of water resources assessment. The availability of scientific information and understanding of the conditions and processes occurring within freshwater ecosystems is an important factor for water managers to systematically evaluate choices and tradeoffs. If the management of freshwater ecosystems is to be improved, Finlayson argues, considerable shifts in decision-making priorities will be required. Where sufficient information and understanding is available, such shifts can be based on improved valuation of ecosystem services. Wider participation by stakeholders is seen as the key to successfully guiding and reorienting such shifts in decision-making and management.

The final paper in this section addresses a more innovative water management approach. Zalewski introduces the concept of Ecohydrology as a management tool through which small-scale technological improvements in water and ecosystem management can be achieved. Ecohydrology focuses on hydrological management at catchment scale to enhance the carrying capacity of ecosystems. The potential for practical application of this framework is demonstrated in approaches to the control of nutrient and pollutant loading in freshwater. This approach is driven by the goal to optimize both water availability and biodiversity conservation, and in this way to arrive at the overall enhancement of ecosystem services for society.

Learning through case studies

The second section of the book presents a series of case studies from the Amazon Region, the Arabian Gulf, East Africa, China, Europe, India and Canada. While the case studies address a diversity of scales and ecosystem conditions, they employ a common approach to the discussion of management challenges. This approach draws upon the MA in its structure, beginning with an evaluation of current conditions and trends of water resources, followed by the development of future scenarios based on projections of water availability and habitat destruction driven by anticipated patterns in population, development patterns, increased demand and environmental change. These scenarios help identify response options to the challenges for the sustainable management of water resources, ecosystems, human well-being and ecosystem services. They also highlight gaps in management capacity.

In their exploration of conditions and trends of water resources, the case study authors consider both per capita water availability in the different locations, as well as water quality and household access to safe water and sanitation. Per capita water availability does not automatically translate into access to water, as can be seen in the case of the water abundant Amazon region, where household access to water remains relatively low. In contrast, household access to water in the arid Arabian Gulf countries is relatively high. These uneven patterns in water resources availability and development often are driven by economic factors and available management capacity. In general, human activities related to economic development are often accompanied by threats to water availability and quality.

The roles that ecosystems play in supplying and protecting water supplies vary amongst the different case studies, which collectively identify a range of environmental processes that are all essential to the provision of water at different ecosystem scales. The chapter by Aragon on the Amazon, focuses on the role of the water cycle in providing and regulating water supplies at the truly regional scale. Environmental flows of water are considered in several chapters, and in particular detail by Chicharo *et al* in their study of the Guardiana Estuary. These are the flows of water that are needed to sustain the functioning of the ecosystem. Many of the case studies examine the capacity of freshwater systems, particularly those with altered flow regimes, to assimilate and remove pollutants. Problems of nutrient loading feature in many of the studies focusing on surface waters, from Canada to China. In many of the locations studied freshwater ecosystems provide the only available means of sewage disposal and treatment.

The capacity of ecosystems to maintain beneficial services for human well-being is under pressure from human-induced changes, such as pollution, water abstraction and man-made alterations to flow-regimes. In some cases, such as those included in the case studies from Portugal, Canada and India, these changes are aggravated by the construction of dams. In the case from China, viewed from the Hengshui Lake, entire fresh water systems have been diverted and relocated. Other processes affecting ecosystems, such as land degradation and deforestation, occurring for example around Lake Victoria and in the Amazon basin, interact with changes in the freshwater system to further decrease the quality of available water.

It is apparent that not all of the trends are negative. A number of the case studies identify positive improvements achieved in recent years through informed management efforts. The creation of biosphere reserves, implementation of environmental monitoring programmes, and application of ecohydrological management techniques, as well as improvements to wastewater treatment programmes, have all brought improvements in water quality and quantity. Many of the notable successes are small-scale management improvements achieved by small communities, such as the community initiatives for water quality management on Lake Victoria or groundwater recharge and village-level water-supply and sanitation improvements in the case from Gujurat.

Capacities must be developed to replicate these successes at the global level; these should include capacities to provide water, to those to measure and understand ecosystem processes supporting water supplies; and therefore to train and educate scientists. In particular, many scientific challenges are identified in which the available understanding of ecosystem processes is not sufficient to enable meaningful assessments of the resource to support decision-making. Challenges remain in all areas, including Canada, where scientists working to gain a better understanding of the assimilative capacity for pollutants in the Saint John River continue to face difficulties due to a lack of baseline data from which to assess the present health of the ecosystem compared to its historical state. However, where scientists have been able to work with local communities to build an understanding of ecosystem processes, innovative management solutions have emerged. For example, in the Guardiana Estuary ecohydrological approaches have helped to reduce the impacts of dam construction on the coastal ecosystem, while in the

Hengshui Lake, China, nutrient removal strategies have succeeded to benefit both the human population and the ecosystem. Examples such as those presented in these case studies, illustrate the potential that participatory scientific approaches to ecohydrology hold for the improvement of water and ecosystem management.

Water and Ecosystems

The main thesis of this book is that conserving ecosystems is important to societies in more ways than one. While ecosystem and biodiversity conservation is of intrinsic value on its own, human society directly and indirectly benefits from this. Healthy ecosystems, achieved through sustainable ecosystem management, are essential for water provisioning to respond to the global water crisis. We have ample evidence, as presented in this book, that it is possible to bring together scientific expertise and community-based wisdom to arrive at successful approaches that improve water supplies to communities through better ecosystem management.

References

Daley, R. and Dowdeswell, E. (2002) *Capacity for Managing Transboundary Coastal Ecosystems: overview of the four case studies.* The Managing Shared Waters Conference. Hamilton, Ontario, Pollution Probe

Ivey, J. L., Smithers, J., Loe, R. C. d. and Kreutzwiser, R. D. (2004) Community Capacity for Adaptation to Climate Induced Water Shortages: Linking Institutional Complexity and Local Actors. *Environmental Management,* **33**(1) pp.36-47

Pirie, R., Loe, R. C. d. and Kreutzwiser, R. (2004) Drought planning and water allocation: an assessment of local capacity in Minnesota. *Journal of Environmental Management,* **73** pp.25-38

UN Millennium Project (2005) Health, dignity and development: what will it take? Task Force on Water and Sanitation, pp. 228 [Online] Available from: http://www.unmillenniumproject.org/reports/reports2.htm [Accessed 18/05/2005]

United Nations Development Programme (2006) Human Development Report 2006 - Beyond scarcity: Power, poverty and the global water crisis, Palgrave Macmillan, New York, USA.

Vorosmarty, C., Leveque, C., Revenga, C., Bos, R., Caudill, C., Chilton, J., Douglas, E. M., Meybeck, M., Prager, D., Balvanera, P., Barker, S., Maas, M., Nilsson, C., Oki, T. and Reidy, C. (2005) Fresh Water. In: (Eds, Hassan, R., Scholes, R. and Ash, N.) Ecosystems and Human Well-being: current state and trends: findings of the Condition and Trends Workding Group. The Millennium Ecosystem Assessment, Vol. 1. Washington, Island Press, pp. 165-207

Section I:

Approaches to Water Resource Management in Diverse Ecosystems

Conditions and Trends of Freshwater Ecosystems and the Challenges to Meet Human Water Needs

Carmen Revenga

The Nature Conservancy, USA

Introduction

The statistics assessing the burden of water-related diseases on human well-being are well known: as of 2002, 1.1 billion people lacked access to safe drinking water and 2.6 billion did not have access to adequate sanitation (WHO, 2004). As a consequence, millions of lives are lost each year. Most of the victims are children under the age of 5, who die from preventable diseases linked to contaminated sources of water (WHO, 2004). With current projections showing an increase in global population of 74.8 million people each year, pressure to meet basic human needs for water is sure to mount. To meet the sanitation Millennium Development Goal alone would mean that an additional 370,000 people would need to gain access to improved sanitation every day until 2015—a substantial level of investment in infrastructure and capacity (WHO, 2004).

Given the challenges to meet people's needs for water, one may ask "why should we care about freshwater biodiversity?" Are species in aquatic systems that important to our own survival and well-being? The answer to this question is undoubtedly: yes. Biodiversity underpins and sustains the functioning of ecosystems and therefore the delivery of services upon which humans depend. Services such as food, clean water, flood protection, and sediment transport to coastal areas. Without species and the ecological process that interconnects them, ecosystem services would be lost, with severe consequences for human well-being. In addition, biodiversity has an intrinsic value. Therefore humans have a responsibility to sustain ecosystems, not only for the survival of other living species, but also for the well-being of future generations.

Because freshwater resources are essential to sustaining life on Earth, proximity to water bodies has been a preference for the establishment of human settlements for millennia. Humans have used rivers for transport and navigation, water supply for domestic uses, for industrial uses, for agriculture and other sources of food, for waste disposal, and for recreation. As a consequence we have heavily altered waterways to fit our needs by building dams, levies, canals, water transfers, and even reversing the direction of flows in entire river stretches. Physical alteration, habitat loss, water withdrawals and diversions, pollution, overharvesting and the introduction of non-native species are widespread across the world and have taken a heavy toll on freshwater biodiversity (Revenga and Kura, 2003).

According to a review by Harrison and Stiassny (1999), the two leading causes for freshwater species imperilment are habitat alteration and the introduction of invasive species. We have altered freshwater ecosystems so much that today freshwater species are, in general, at higher risk of extinction than those in forests, grasslands, and coastal ecosystems (WRI et al., 2000). For North America alone, a comparatively data-rich country, the projected extinction rate for freshwater fauna is five times greater than that for terrestrial fauna—a rate comparable to the species loss in tropical rainforest (Ricciardi and Rasmussen, 1999). Despite these warning signs and projections, the level of assessment and knowledge of the status and trends of freshwater species are still very poor. Overall, the damage done to freshwater ecosystems is inevitably underestimated because extinctions go unseen where the species lost have not yet been taxonomically classified. Unfortunately, increasing human populations and achieving the sustainable development targets set forth in 2002 will place even higher demands on the already stressed freshwater ecosystems.

Conflicting Commitments

One of the major challenges that human society faces today is achieving those Millennium Development Goals (MDGs) that relate to the use of water resources while at the same time sustaining functioning freshwater ecosystems and reducing the rate of biodiversity loss. These goals include halving the proportion of people that suffer from hunger and the proportion of people without access to safe drinking water and adequate sanitation by 2015. On one hand, governments have agreed to increase crop production and water supply facilities, as well as water delivery infrastructure to achieve the MDGs. On the other hand, they have made these pledges while committing to reducing the rate of loss of biodiversity by 2010 and to ensure environmental sustainability.

Meeting the targets for a stable global climate through renewable energy development places an additional pressure on freshwater ecosystems and species. Hydropower is the leading alternative energy source if compared to energy derived from fossil fuel. However, hydroelectricity production requires the construction of dams, which in turn will fragment and impact riverine ecosystems and their dependent species, unless they are built and operated to account for needed environmental flows.

Each of the MDG or climate stabilization targets, presents a separate challenge. To achieve all of them without accelerating the rate of biodiversity loss in freshwater ecosystems is going to require much effort – not only in terms of data, information and resources, but political commitment, and drastic changes in how we manage and allocate water resources. Given these pressures, assessing the condition and rates of change of freshwater species and habitats is of critical importance for preserving the integrity of these ecosystems and the goods and services we derive from them.

Freshwaters are Unique Ecosystems

Freshwater ecosystems are highly rich in species richness and endemism. To date, there are approximately 44,000 scientifically described freshwater species (Reaka-Kudla, 1997). While in absolute terms this figure is not as high as for other realms (i.e., marine), when considering the relatively small portion of the planet these species occupy—less that 1% of the Earth's surface—freshwater ecosystems are highly rich in species (Reaka-Kudla, 1997). The number of freshwater species scientifically described is also increasing rapidly as more and more species are identified and catalogued. In fact, about 200 new freshwater fish species are described each year (Lundberg et al., 2000). Endemism is also unusually high, with, for example, 632 endemic animal species recorded in Lake Tanganyika, and an estimated 1,800 species of fish endemic to the Amazon River basin (WRI et al., 2003).

Freshwater ecosystems are also very resilient, with examples of species refugia found in highly altered river systems (Lévêque per. comm., 2004). This resiliency, however, is not infinite. We know that there are thresholds, which, once crossed, can put entire ecosystems at risk. These risks entail severe consequences for human well-being and biodiversity. One of the best documented examples of ecosystem degradation is the case of the Aral Sea, where biodiversity, human health and livelihoods have paid a heavy toll for the narrow interests of the agricultural community. From 1960 to 1990, irrigated land in the Aral Sea region almost doubled. The irrigated area was extended from 4.5 million to 7 million hectares, with 90% of the available water from the ecosystem going to meet irrigation needs (UNESCO, 2000). The excessive water diversion caused reduced flows in the two main rivers feeding into the Aral Sea: the Syr Darya and the Amu Darya. As a consequence, the water volume in the Aral Sea basin has been reduced by 75% since 1960. The once rich fisheries have disappeared as well as many of the aquatic and terrestrial species dependent on the Sea's wetlands and river deltas (UNESCO 2000; Postel, 1999). The consequences have also included severe impacts on human well-being: jobs and livelihoods have been lost due to loss of fisheries and navigation; declines in agricultural productivity have occurred in the deltas because of increase water salinity; and health problems such as increased incidence of pulmonary and other diseases have resulted from a of lack of drinking water and exposure to chemicals from agricultural runoff in the dry seabed (Postel, 1999).

Experience has shown that it is much more expensive to restore ecosystems than to prevent their degradation. Even with substantial investment, reversal of degradation to pre-modification conditions is not always possible to achieve and can take many years. This is particularly true when dealing with species changes. On the one hand, once a species is lost, it cannot be recovered and remains permanently extinct. On the other hand, once some invasive species are established, they are costly to remove or even keep in check. The everglades restoration plan, for example, is estimated to cost $7.8 billion and will take more than 30 years to complete (CERP, 2005).

At the same time, in many instances, it is hard to measure in economic terms the impact that altered and degraded ecosystems have on human well-being. Usually there is a time lag between the cause of degradation and the effect of such alterations on ecosystems, species populations, and human well-being. This time lag makes it difficult to assess the impacts of proposed development plans on ecosystems and to estimate the costs and benefits to local communities. What we do know is that the poor suffer the most when ecosystems are degraded, as their coping capacity is lower. Poor rural populations tend to have fewer development alternatives and tend to depend directly on ecosystems for their livelihood. Inland fisheries, for example, are in many instances the employment of last resort when other economic opportunities are lost. In the developing world, many farmers supplement their incomes with fishing. For instance, in Laos, 70% of farm households augment their food supply and income by fishing (Sverdrup-Jensen, 2002). In the Mekong River basin, an estimated 40 million farmers are also engaged, at least seasonally, in fishing activities (Sverdrup-Jensen, 2002).

Given the importance of these ecosystems in sustaining human well-being, it is surprising how little we know about their changing condition, their dependent species, or the roles that these species play in sustaining ecological functions. This lack of knowledge and information, unfortunately translates into a paucity of indicators that policy makers can use to evaluate tradeoffs when assessing management and development plans for water resources and freshwater ecosystems. The following section briefly examines our current knowledge of freshwater biodiversity and suggests some indicators that can be used to inform policy decisions.

What do we know About Trends in Freshwater Species?

Data on the condition and trends of freshwater species are, for the most part poor at the global level, although some countries have better inventories and indicators of change in freshwater species (e.g., Australia, Canada, and United States). This does not mean, however, that there are no data available. There are considerable data on freshwater species and populations, but these are not necessarily accessible. For example, many countries have large inventories of freshwater species in their museum and university collections, but these data are rarely centrally located or electronically archived. Often country-level information is in national languages that are not accessible to the larger scientific community. International reviews of research tend to rely on English, and to some extent French and Spanish literature (Revenga and Kura, 2003).

More fundamentally, large numbers of specimens have never been catalogued. Although historic museum records cannot alone be used to monitor the status of populations and species into the future, they can help to establish baseline species distributions against which to evaluate current and future conditions. Knowledge is particularly poor for lower taxonomic groups (aquatic plants and invertebrates). The better studied groups are amphibians, water birds, and fish. However, for this last group of species some continents (e.g., North America, Africa, Europe, Australia) are much better studied than others (South America, Asia, Oceania). Fish inventories for South America and Asia are still limited with some regions not yet sampled.

Amongst the available species data and indicators, there is a considerable bias toward temperate regions, especially North America. The coverage of freshwater species in the IUCN 2003 *Red List of Threatened Animals,* for example (see Table 1), is still minimal for most taxa. And while all the mammals have been assessed for threat status, many of the freshwater-dependent mammals are categorized as data deficient, meaning that there is not sufficient information on their populations to assign a threat category. Fortunately, IUCN's Species Survival Commission and other organizations are currently conducting a full assessment and mapping exercise for all mammals and reptiles of the world, and have just finalized a Global Amphibian Assessment (see http://www.globalamphibians.org/).

Table 1: Coverage of Freshwater Species in the 2003 IUCN Red List of Threatened Animals.

Taxon (Freshwater Dependant)	Estimated No. of Freshwater dependent species or subspecies	Est. % Assessed for the IUCN Red List 2003
Plants		< 0.1%
Insects	1000 – 10,000	< 0.1%
Molluscs	6,000	< 15%
Crustaceans	??	< 5%
Fish	>11,000	< 10%
Reptiles	266	< 15%
Amphibians	>5,500	100%
Water Birds	868	100%
Mammals	> 135	100%

Source: IUCN/SSC 2004

The IUCN *Red List* on the threatened status of plants and animals is one of the most widely used indicators for assessing the condition of ecosystems and their biodiversity. The list is developed according to a harmonized category and criteria classification, where all contributing experts follow the same methodology and guidelines. Because of this, it is the best source of information, at the global level, on the conservation status of plants and animals. This system is designed to determine the relative risk of extinction to species. The main purpose is to catalogue and highlight those taxa that are facing a higher risk of global extinction (i.e. those listed as Critically Endangered, Endangered and Vulnerable). It is important to note, however, that there is a considerable geographic bias towards North America in the *Red List* assessment for freshwater species (see Table 2). This bias is probably driven by data availability, combined with the high level of knowledge, and research capacity in this region that is represented among the IUCN/SSC expert network.

Table 2: Regional Representation in the Coverage of Freshwater Species in the 2003 IUCN Red List of Threatened Animals.

World Regions	No. Species Listed	(Total 3010)
Antarctic	3	(0.1%)
Caribbean Islands	60	(2%)
East Asia	160	(5.3%)
Europe	348	(11.6%)
Mesoamerica	207	(6.9%)
North Africa	23	(0.8%)
North America	892	(29.6%)
North Asia	117	(3.9%)
Oceania	408	(13.6%)
South and South East Asia	347	(11.5%)
South America	191	(6.3%)
Sub-Saharan Africa	414	(13.8%)
West and Central Asia	148	(4.9%)

Source: IUCN/SSC 2004

Freshwater Species are in Decline and Increasingly Threatened

Numerous studies have shown that freshwater species are increasingly threatened in many parts of the world (WRI et al., 2000; Gleick et al., 2001; Revenga and Kura, 2003). The United States is one of the few countries to that conducts relatively comprehensive assessments of the conservation status of freshwater molluscs and crustaceans. In this country, one-half of the known crayfish species and two-thirds of freshwater molluscs are at risk of extinction (Master et al., 1998), with severe declines in their populations recorded in recent years. Furthermore, of the freshwater molluscs identified, at least 1 in 10 is likely to have already gone extinct (Master et al., 1998).

In terms of freshwater fish, it is estimated that more than 20% of the 10,000 described freshwater fish are threatened, endangered or have become extinct since the 1950s (Moyle and Leidy, 1992). Even commercial fisheries that used to be plentiful show marked declining trends. To estimate the declines in freshwater fisheries catch is difficult, given that the catch is greatly under-reported- by a factor of 3-4 (FAO, 1999). The last major assessment of inland fisheries by the Food and Agriculture Organization of the United Nations (1999) reported that most inland capture fisheries that rely on natural reproduction of the stocks are either already overfished, or are being fished at their biological limit. The principal factors threatening inland capture fisheries are fish habitat loss and environmental degradation. In fact, many inland fisheries are being sustained only through fishery enhancements (e.g., stocking, introduced species, etc).

In terms of amphibians, the Global Amphibian Assessment completed in 2004, shows a declining trend. Assessment findings suggest that the rate of decline in the conservation status of freshwater amphibians is far worse than that of terrestrial species. Nearly one-third (32%) of the world's amphibian species were found to be threatened, representing 1,856 species (IUCN et al., 2004). By comparison, just 12% of all bird species and 23% of all mammal species are threatened.

Data on freshwater reptiles, namely freshwater turtles and crocodilians (i.e., crocodiles, caimans, and gharials) also show similar trends. According to the IUCN/SSC Tortoise and Freshwater Turtle Specialist Group and the Asian Turtle Trade Working Group, of the 90 species of Asian freshwater turtles and tortoises, 74% are considered threatened. Over half of Asian freshwater turtle and tortoise species are endangered, including 18 critically endangered species, and one that is already extinct: the Yunnan box turtle (*Cuora yunnanensis*) (van Dijk et al., 2000). The number of critically endangered freshwater turtles has more than doubled since the late 1990s (van Dijk et al. 2000). Much of the threat has come from overexploitation and the illegal trade in Asia.

The status of crocodilians presents a similar pattern, particularly in Asia. Of the 17 freshwater-restricted crocodilian species, 4 are listed as Critically Endangered (3 of which are in Asia), 2 as Endangered and 2 as Vulnerable (IUCN, 2003). The most critically-endangered is the Chinese alligator (*Alligator sinensis*). The major threats to crocodilians worldwide are: habitat loss and degradation caused by pollution, drainage and conversion of wetlands, deforestation, and overexploitation (Revenga and Kura 2003; IUCN, 2003).

Water birds, and particularly migratory water birds, are relatively well studied, with time series data being available in North America and north-west Europe for about 30 years. The Red List Index for birds (Figure 1) developed by BirdLife International and IUCN, provides the best estimate of net changes over time to the overall threat status of the world's birds. With the exception of marine bird species, freshwater-dependent bird species show the sharpest and continuous decline over time. The recent sharp decline in marine birds is predominantly caused by an increase in long-line fisheries, which have high bird bycatch rates, particularly albatrosses and large petrels (Butchart et al., 2004).

Figure 1: Red List Index for Birds.

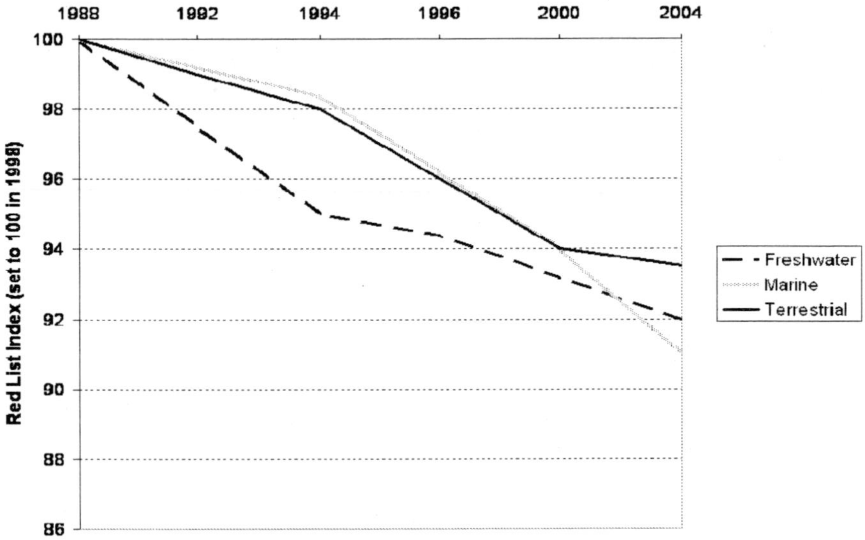

Source: Modified from Butchart et al.,2004.

Although all terrestrial mammals depend on freshwater for their survival and many feed and drink in rivers and lakes, only a small number of mammals are considered inland aquatic or semi-aquatic mammals. These mammals spend a considerable amount of time in freshwater and usually live in riparian vegetation close to water bodies such as rivers, lakes, lagoons, and ponds or in marshes and swamps, although they may forage and sleep on land. Information on these species is fragmentary; some species have been better studied and sampled than others. For instance, freshwater cetaceans (e.g., dolphins), probably because of their endangerment, are better known than freshwater insectivores (e.g., Pyrinean desman) or mustelids (e.g., otter civet).

Where information is available on freshwater mammals, their conservation status is not very promising. Freshwater cetaceans are among the most threatened mammals in the world. There are 5 species of river dolphins and one species of freshwater porpoise living in large river systems of Asia and South America. Populations of these species have declined rapidly in recent years and much of their habitat has been degraded (Revenga and Kura, 2003). The Asian species are highly threatened. There are also several species and subspecies of freshwater seals in lakes in Russia and Europe. Table 3 presents the distributions of these species and their level of threat by basin.

Table 3: Species of Freshwater Cetaceans and Pinnipeds and their Conservation Status.

Basin Name	Species Common Name	Conservation Status*	Population Numbers
Yangtze	Yangtze River dolphin (or baiji)	CR	< 50
Yangtze	Finless porpoise	EN	<2,700 (as of 1991)
Ganges/ Brahmaputra	Ganges River dolphin (or susu)	EN	<1,200
Indus	Indus River dolphin (or bhulan)	EN	<1,000
Irrawaddy/Mekong	Irrawaddy River dolphin	DD	?
Mahakam (Kalimantan)	Irrawaddy River dolphin subpop.	CR	<30-50 (as of 2000)
Amazon/Orinoco	Amazon River dolphin	VU	?
Lake Baikal (Russia)	Lake Baikal seal	LR/nt	60,000
Lake Ladoga (Russia)	Lake Ladoga seal (subsp.)	VU	2,000
Lake Saimaa (Finland)	Lake Saimaa seal (subsp.)	EN	220-250

*IUCN Conservation Status: CR: Critically Endangered, EN: Endangered, LR/nt: Lower Risk near threatened, VU: Vulnerable, DD: Data Deficient.

Source: Data on cetaceans compiled by Revenga and Kura 2003 from multiple sources; Seal Conservation Society, 2001.

Another indicator used to look at trends on actual populations of species is WWF's Living Planet Index (LPI). The LPI provides a measure of the trend in vertebrate species populations based on data assessing approximately 3,000 population trends for more than 1,100 species around the world, of which 323 are freshwater species. The index shows that freshwater populations declined to a greater extent between 1970 and 2000 than those in marine and terrestrial systems (Loh and Wackernagel, 2004). The freshwater species index (Figure 2), shows a decline of approximately 50 per cent from 1970 to 2000 compare to 30% decline for marine and terrestrial species. Information from the Nearctic and Palearctic realms predominate in the freshwater LPI, but the method of calculation compensates for this to some extent by giving equal weight to data from all realms.

Figure 2: Living Planet Index.

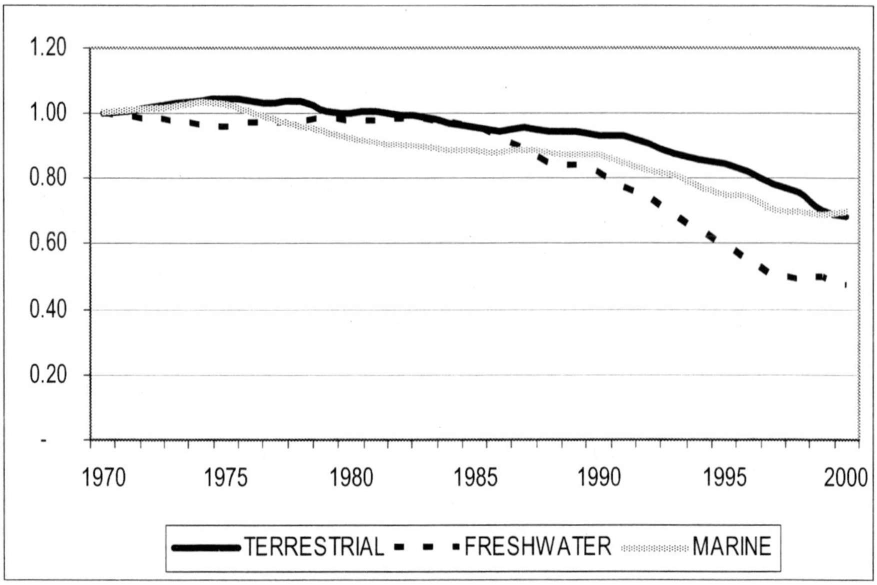

Source: Modified from Loh and Wackernagel, 2004.

Freshwater Habitats and Water Resources: What do we know?

With regards to our knowledge of freshwater ecosystems, we are not much better off than we are regarding freshwater species. Finlayson & Spiers (1999) carried out an assessment of our knowledge of the extent, distribution, and change of wetlands and concluded that based on existing information it is not possible to reliably estimate the total extent of wetlands at a global scale. Their assessment found that of the 206 countries assessed: 7% had adequate or good national inventory coverage, 69% only had partial coverage, and 24% had little or no national wetland inventory. This is due in part to differences in definitions, as well as difficulties in delineating and mapping habitats with variable boundaries due to fluctuations in water levels.

Even with the use of remote sensing, the wide range in the sizes and types of wetlands and the problem of combining hydrologic and vegetation characteristics to define wetlands make it difficult to produce a global, economical, and high-resolution data set with existing sensors, except for large water bodies in arid and semi-arid regions of the world. Larger wetlands and inland seas have been mapped, but smaller habitats, seasonal and intermittently flooded wetlands and many flooded forests that are critical from a biodiversity standpoint are not well mapped or even delineated in many parts of the world.

Lakes have been mapped reasonably well, although issues of scale also occur. However, change in lake extent or condition over time is not regularly monitored except for a few large lakes, such as the North American Great Lakes. The International Lake Environment Committee (ILEC) maintains a database of over 500 lakes worldwide, with some physiographic, biological and socio-economic information (Kurata 1994; ILEC 2002), and while data collected highlight major problem areas that are widespread among lakes and reservoirs, such as lowering of the water level, siltation, acidification, chemical contamination, eutrophication, salinization, and the introduction of exotic species (Kira, 1997; Jørgensen et al., 2001), the information is questionnaire-based, largely descriptive, often incomplete, and not regularly updated to be used in long-term monitoring.

For river systems, measures of condition require both maps of rivers and delineations of catchments, as well as information on discharge. Global inventories of major river systems, including data on drainage area, length, and average runoff are available (e.g. Baumgartner and Reichel, 1975; Shiklomanov, 1997; WRI et al., 2000), but they also suffer from differences in definitions of the extent of a river system and the time period or location for the measurement of discharge (Revenga and Kura, 2003). Calculating drainage area, for example, requires the delineation of each catchment, data on river networks, and topographic maps. Modeled discharge data for the world's rivers can be used with the global drainage map, but measured discharge data are far preferable. The World Meteorological Organization's (WMO) Global Runoff Data Centre compiles and maintains a database of observed river discharge data from gauging stations worldwide. Although this is the best global database currently available, the number of operating stations has significantly declined since 1980s—meaning the discharge data for many rivers have not been updated in the last two decades – this is particularly true for Africa. Better and more reliable information on actual stream and river discharge, and the amount of water withdrawn and consumed at the river basin level, would increase our ability to manage freshwater ecosystems more efficiently and set conservation measures for ecosystems and species. However, much effort and financial commitment would have to be made to restore hydrological stations.

Chemical water quality, including physical parameters such as temperature, has been the most widely used tool to evaluate physical habitat quality in freshwaters. Most national governments have some sort of chemical water quality monitoring program in place, although in many developing countries geographical coverage is limited and quality assurance procedures may be less than satisfactory. Global chemical water quality data is collected through the UN GEMS (Global Environmental Monitoring System) program which collects data from 106 countries (www.gemswater.org/global_network/index-e.html). But the actual data contributed differs between countries, with many monitoring stations collecting a few variables, from which it is hard to extrapolate water quality parameters to entire basins. In any case, the value of chemical water quality data as an indicator of freshwater biodiversity is limited.

For groundwater, our knowledge and monitoring capacity is even worse – with very few countries comprehensively monitoring water quality or level. Part of the problem in monitoring and data collection is that our Institutional structures are not set up to collect, share, and analyze information in a way that favors integration and cross-sectoral management. In most countries water resources are managed in a piecemeal fashion with a particular sector, such as the ministries of transportation, for example, having jurisdiction over managing main river channels while other departments are charged with managing and monitoring other parts of the watershed. Analyses are done at scales that are not always appropriate to evaluate tradeoffs. And finally, the communities that are most affected by changes in water resources allocation or management are rarely consulted.

Given the paucity of data, what can we measure to achieve global monitoring of the extent and condition of freshwater ecosystems and their dependent species? We can infer condition and change in freshwater habitats by using global and continental geospatial indicators to assess pressures. For example, by using data on the extent of agriculture in a watershed, or the size and location of dams, we can draw some conclusions about the relative degree of alteration or stress affecting a system. These geospatial indicators are often called proxies or surrogates, because they are indicators of current threat and give only indirect information about actual ecological integrity. We develop these indicators at larger spatial scales using geospatial datasets and a Geographic Information System (GIS). GIS allows us to analyze the spatial relationships between anthropogenic activities and freshwater systems. Innumerable analyses can be run using geospatial data; the challenge is to select the most appropriate tools based on our best ecological knowledge and available data, and then to undertake a judicious interpretation of the results. The examples presented here focus on threat assessments over large areas with limited data available, as is the case in much of the developing world. While global-scale assessments of are necessarily coarse given the available input data, they serve to illustrate spatial patterns that can be used as tools for raising awareness and informing decision makers.

Indicator of River Fragmentation and Flow Regulation

This indicator represents the interruption of a river's natural flow by dams and provides a measure of the degree to which rivers have been modified by humans. The impoundment of main channels, the presence of dams on major tributaries, the storage volume of reservoirs, and the overall reduction of discharge are all considered to be substantial threats to the integrity of river systems (Nilsson et al., 2005). Strongly affected rivers as those where only 25km of the main river channel is left without dams and unaffected rivers are those with no dams in the main channel and if one of the tributaries has a dam, the flow is not regulated by more than 2% (Figure 3).

The results of the fragmentation and flow regulation indicator analysis, demonstrated that of the 292 large river systems assessed, 60 percent are highly or moderately fragmented by dams (Nilsson et al., 2005). The only remaining large free-flowing rivers in the world are found in the tundra regions of North America and Russia, and in smaller coastal basins in Africa and Latin America. It should be noted, however, that considerable parts of some of the large rivers in the tropics, such as the Amazon, the Orinoco, and the Congo, would be classified as unaffected rivers if an analysis at the subbasin level were done. Overall strongly or moderately fragmented systems affect approximately 85 percent of the water volume in these rivers (Nilsson et al., 2005). All river systems with parts of their basins in arid areas or that have internal drainage systems are highly fragmented.

Figure 3: Fragmentation and Flow Regulation Indicator.

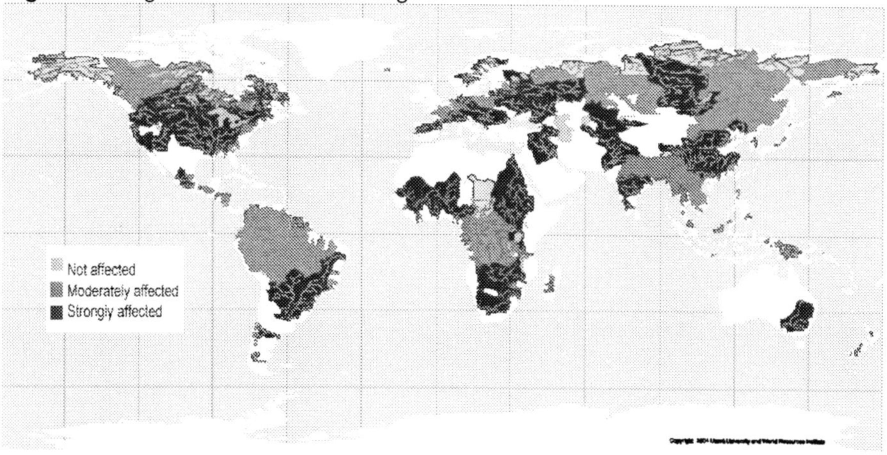

Source: Nilsson et al., 2005.

Large Dams under Construction

Another indicator that can be used to look at future threats to freshwater species is the number of large dams under construction in each river basin. The environmental impacts associated with large dams are well documented (WCD, 2000). These impacts vary in extent and gravity, but it should be stressed that the impacts of large dams are often basin-wide. Dams disconnect rivers from their floodplains, interrupt fish migrations, change the temperature and velocity of in-stream flow, and flood essential feeding and breeding habitats for many aquatic and terrestrial species.

The indicator presented in Figure 4 shows that the Yangtze Basin in China has the highest number of large dams under construction (46), followed by the La Plata basin in Brazil and Argentina with 27 large dams under construction, and the Tigris and Euphrates basin in Turkey, Syria and Iraq with 26.

Figure 4: Large Dams under Construction, 2003.

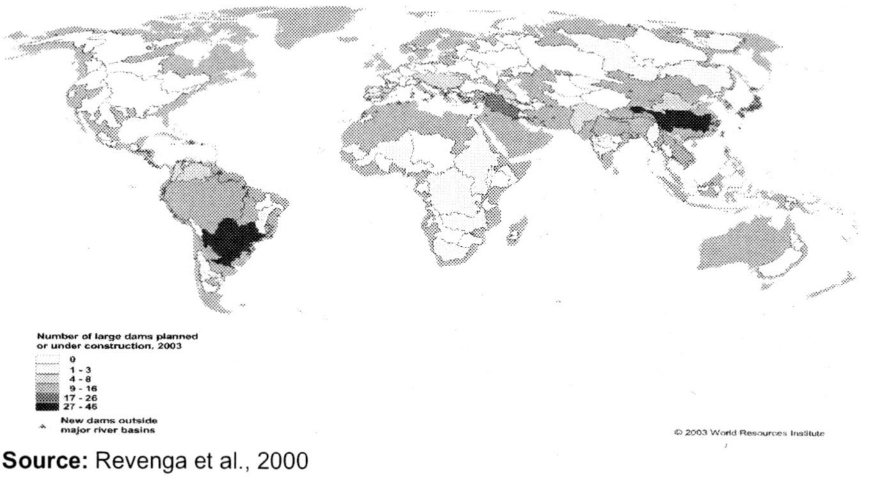

Source: Revenga et al., 2000

Water Stress

Water stress is another indicator of relevance to aquatic species as well as to human well-being. Experts define areas where per capita water supply drops below 1,700 m^3 per year as experiencing "water stress" —a situation in which disruptive water shortages can frequently occur. In areas where annual water supplies drop below 1,000 m^3 per person per year, the consequences can be more severe and lead to problems with food production and economic development unless the region is wealthy enough to apply new technologies for water use, conservation, or reuse.

This indicator analysis projects that by 2025, assuming current consumption patterns continue, at least 3.5 billion people —or 48 percent of the world's projected population —will live in water-stressed river basins (Revenga et al., 2000). Of these, 2.4 billion will live under high water stress conditions. This per capita water supply calculation, however, does not take into account the coping capabilities of different countries to deal with water shortages (Figure 5). For example, high-income countries that are water scarce may be able to cope to some degree with water shortages by investing in desalination or reclaimed wastewater. The study also discounts the use of fossil water sources because such use is unsustainable in the long term.

Figure 5: Water Stress Projections for 2025.

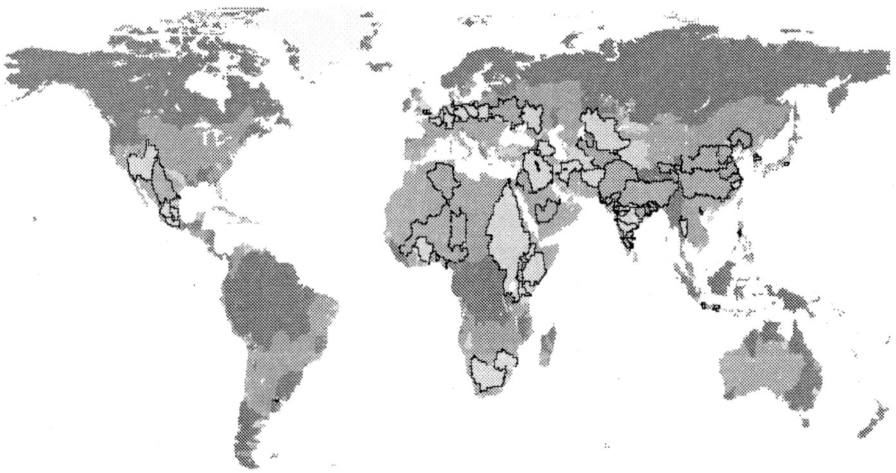

Source: Revenga et al., 2000.

Those basins outlined in the map represented watersheds under water scarcity situations in 2025 and where the projected population is expected to be higher than 10 million. These basins include, among others, the Volta, Farah, Nile, Tigris and Euphrates, Narmada, Jubba, Godavari, Indus, Tapti, Syr Darya, Orange, Limpopo, Huang He, Seine, Balsas, and the Rio Grande and the Colorado River basins in the United States.

Environmental Water Scarcity

Most of the water scarcity assessments done at the global and regional scales assume all runoff is available for humans, which is not the case if we need to sustain functioning ecosystems. A first attempt to include environmental water requirements into global water scarcity assessments shows a grimmer picture (Figure 6) – particularly in arid, semi-arid and dry subhumid areas of the world. When ecosystem's water requirements are taken into account, more basins show a higher degree of water stress. This is not surprising as, if water is reserved for the environmental purposes, its availability for other human uses naturally decreases. This seemly straightforward fact is hardly taken into account in most current water scarcity assessments and projections. And, given that many livelihoods, especially those of the poor, depend on productive freshwater dependent ecosystems, current global water assessments overestimate the amount of water directly available for people.

The indicator shown in Figure 6 represents water withdrawal as a proportion of water available for human use. In this context, the basins with higher proportional extraction levels are those where humans are at a higher risk of water stress if water allocations are made for the maintenance of freshwater dependent ecosystems in moderately modified conditions.

Figure 6: Environmental Water Scarcity.

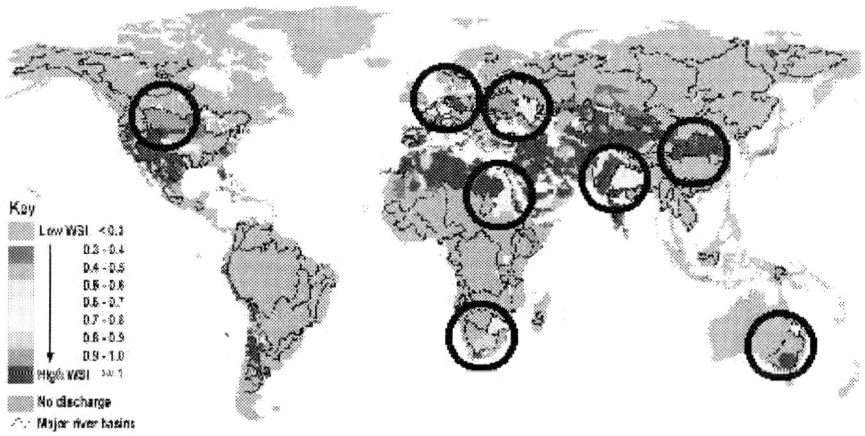

Source: Smakhtin et al., 2004.

The circled basins in Figure 6 are a few examples where overabstraction of water is causing problems to the ecosystems and to the people that depend on their environmental services. *The Murray–Darling Basin* in Australia, with a water stress indicator greater than 1, is an example of an environmentally water scarce basin. This basin, the largest in Australia with just over 1 million km^2, has a highly uneven distribution of flow (both spatially and in time). Throughout Australia's history, the rivers in the basin have been severely modified and regulated. The main economic activity, which uses 95 percent of the total water withdrawal in the basin, is irrigated agriculture. This sustained over-abstraction of water has negatively impacted agricultural production and has caused severe environmental problems in the system. Impacts include high salinity levels that affect soil productivity, massive algae blooms, nutrient pollution, and the consequent loss of native species, floodplain areas, and wetlands.

Implications for Water Management at a Basin Scale

It is at the scale of the river basin that monitoring becomes most relevant, as this is the scale at which interventions begin to have the potential to make a difference for freshwater ecosystems. Perhaps somewhat surprisingly, better quality data at the higher resolution of the river basins are not necessarily more readily available. For example, water use may be estimated for a country or large river basin through summing annual industrial, household, and agricultural water withdrawals across the entire area, but characterizing the spatio-temporal distribution of water withdrawals within a given river basin would require locational and other data that rarely exist.

On the other hand, at the river basin scale high-resolution remote sensing imagery analyses may be feasible, both from a financial and workload perspective, due to the smaller region of analysis. Within a relatively small study area, gauging stations and dam locations may be reliably linked to river stretches; digital elevation models can be improved to reproduce the actual river system; rivers, lakes, and wetlands can be aligned and topologically connected; and species occurrences can be mapped to particular streams, river reaches, or other freshwater systems.

This example of the monitoring in the Upper-Vaal catchment in South Africa, illustrates how a basin-level analysis can inform management. South Africa at present is one of the best examples of a national concerted effort to monitor, assess, and conserve the country's aquatic biodiversity. One of its most prominent and innovative pieces of legislation is the National Water Act of 1998. The fundamental guiding principles of South Africa's Water Act are sustainability and equity in the "protection, use, development, conservation, management and control of water resources." The Act also establishes that the national government, acting through the Minister, is responsible for the achievement of these fundamental principles in accordance with the Constitutional mandate for water reform (Department of Water Affairs, Rep. of South Africa, 1998).

One of the most progressive aspects of the National Water Act is the establishment of the *Reserve*, which consists of two parts: the basic human needs reserve and the ecological reserve. The basic human needs reserve provides for "the essential needs of individuals served by the water resource in question and includes water for drinking, for food preparation and for personal hygiene". The ecological reserve relates to both the quantity and quality of the water required to protect the aquatic ecosystems of the water resource, and will vary depending on the class of the resource. The government is required to determine the *Reserve* for all or part of any significant water resource.

Since the enactment of the Act, the South African Department of Water Affairs and Forestry has engaged in a complex program of implementation—of which the River Health Programme (RHP) is a component. The RHP's main purpose is to serve as a source of information regarding the overall ecological status of river ecosystems in South Africa in order to support the rational management of these natural resources. It uses biological indices (e.g. fish communities, riparian vegetation, aquatic invertebrate fauna) to assess the condition or health of river systems, and to characterize the response of the aquatic environment to multiple disturbances. The rationale behind the use of these indices is that by monitoring and assessing the condition of river system and other aquatic resources, it is possible to identify and report emerging problems regarding aquatic ecosystems, and take corrective action.

The Upper Vaal catchment, in the Free State Province is as an example of the implementation of the RHP in South Africa. Some of the indices used to assess the river health, and shown for this example, include:

- Index of Habitat Integrity (IHI), which assesses the effect of disturbances on instream and riparian zone habitat;
- South African Scoring System (SASS), which assesses the invertebrate families found at each sampling site; and
- Fish Assemblage Integrity Index (FAII), which is an expression of the degree to which a fish assemblage deviates from its undisturbed condition

The Habitat Integrity Index, for example, involves the flying over each river and identifying impacts and changes throughout the river course. This provides scientists with a holistic perspective of the river. It is also an ideal opportunity to select monitoring and sampling sites, as one can identify pollution sources and other stressors. These indices, are then applied and tailored to local conditions by Provincial authorities and local specialists. Within a catchment, each river is assigned a "desired health" status, depending on the potential to deliver a range of goods and services (RHP, 2003). For instance, a "fair" or "poor" river may have lost its potential for most of the uses such as tourism and recreation, conservation, etc. A fair river, however, may still provide water for irrigation. The desired state of each river is determined by stakeholder participation and taking into account socio-economic variables in that particular basin. Trade-offs are evaluated within this context keeping in mind the ultimate goal of sustainable use of water resources. Once the desired state of each river has been determined, monitoring and assessment of the current health status of each river using the above indices, is carried out as part of the biological monitoring program. Figure 7 shows the results of these two processes for the Upper Vaal Basin.

Figure 7: Monitoring and Assessment Indicators for the Upper-Vaal River Basin,

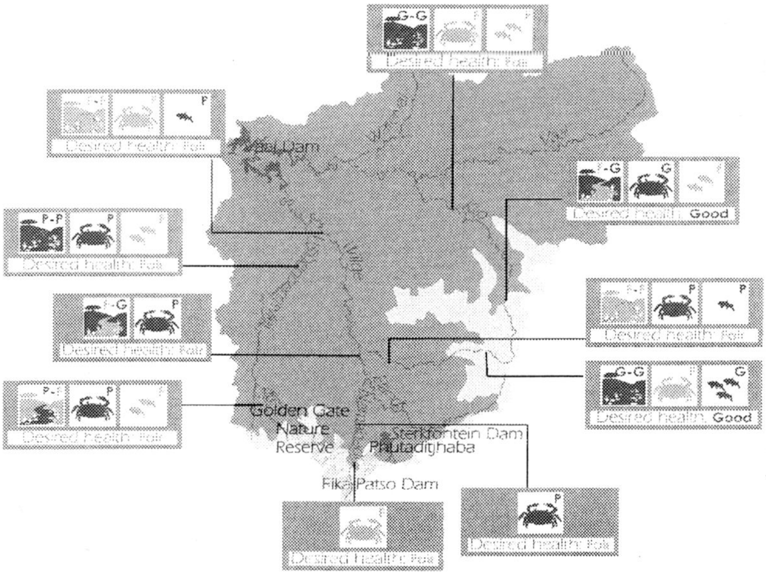

Source: Free State Province, River Health Programme 2004.

Conclusions and Recommendations

As these examples show, monitoring and assessment indicators of the condition and change in freshwater ecosystems can help identify which pressures (pollution, abstraction, damming, over utilisation, etc.) are affecting different river stretches and to what extent species are being impacted. Based on these findings, management strategies to utilize water resources to meet the needs of people and ecosystem can then be developed.

But in order to carry out monitoring and assessment programs there needs to be a concerted data collection effort to establish baselines against which we can monitor change over time. Baseline information is needed at the basin scale in most areas related to water resources, from species richness and endemism inventories, to data on pressures on freshwater ecosystems such as the location of dams, water withdrawals, socio-economic data, and water quality.

In addition to data collection, it is critical to involve local communities particularly the poor in decision making processes, as well as to develop ways to value ecosystem services in economic and non-economic terms. Without these values, it will be hard to incorporate ecosystem services into cost benefit analyses on a regular basis and highlight the dependence of certain communities on functioning ecosystems.

Another crucial step in managing freshwater resources more sustainably is to move away from the traditional sectoral approach to water management that favors single objectives with high economic returns. For example, the conservation community has to increasingly engage the agriculture and irrigation community as well as the infrastructure community (e.g., dam operators and builders) to find common management practices that deliver water to people while sustaining functional ecosystems.

Finally, we need to mainstream the notion of ecosystems as legitimate users of water. This will require incorporating the notion of ecosystem water requirements into water resources assessments and water allocations, as well as valuing water appropriately. In my view, long-term sustainability in water management will be achieved when we see a real shift from the traditional belief that "water that reaches the sea is water wasted" to an ecosystem-based belief that sustains the notion that water that remains in the river is an integral part of water management.

References:

Baumgartner, A. and E. Reichel. (1975) *The World Water Balance: Mean Annual Global, Continental, and Maritime Precipitation, Evaporation, and Runoff*, Amsterdam, The Netherlands: Elsevier Science Publishers.

Butchart, S.H.M., A.J. Stattersfield, L.A. Bennun, S.M. Shutes, H.R. Akçakaya, J.E.M. Baillie, S.N. Stuart, C. Hilton-Taylor and G.M. Mace. (2004) "Measuring global trends in the status of biodiversity: Red list indices for birds", *(Public Library of Science) PLoS Biology* 2(12).

CERP (2005) *The Comprehensive Everglades Restoration Plan (CERP)*, available from http://www.evergladesplan.org/about/rest_plan.cfm.

Department of Water Affairs, Rep. of South Africa. (1998) *South Africa's National Water Act (Act No. 36),* available from http://www-dwaf.pwv.gov.za/Documents/Legislature/nw_act/NWA.doc.

FAO (Food and Agriculture Organization of the United Nations). (1999) *Review of the state of world fishery resources: inland fisheries.* FAO Fisheries Circular No. 942. Rome, Italy: FAO Fisheries Department.

Finlayson, C.M. and A.G. Spiers (eds). (1999) *Global review of wetland resources and priorities for wetland inventory. 2nd edition*, Wageningen, The Netherlands and Jabiru, Australia: Wetlands International and Environmental Research Institute of the Supervising Scientists.

Gleick, P.H., A. Singh, and H. Shi. (2001) Threats to the World's Freshwater Resources. A Report of the Pacific Institute for Studies in Development, Environment, and Security, Oakland, California, in cooperation with the United Nations Environment Programme.

Harrison, I.J. and M.J. Stiassny (1999) "The quiet crisis: a preliminary listing of the freshwater fishes of the world that are extinct or 'Missing in Action' ", in R.D.E. MacPhee, ed., *Extinctions in near time: causes, contexts, and consequences.* New York: Kluwer Academic/Plenum Publishers, pp. 271–332.

ILEC (International Lake Environment Committee). (2002) available from http://www.ilec.or.jp/database/database.html.

IUCN/Species Survival Commission. (2004) *Red List of Threatened Species.* IUCN – The World Conservation Union, Gland, Switzerland and Cambridge, U.K.

IUCN (The World Conservation Union). (2003) *IUCN Red List of Threatened Species. The IUCN Species Survival Commission*, available from http://www.redlist.org.

IUCN, Conservation International, and NatureServe. (2004) *Global Amphibian Assessment*, available from http://www.globalamphibians.org/.

Jørgensen, S.E., R. de Bernardi, T. Ballatore, and V. Muhandiki. (2001) Lake Watch 25: changes in the world's lakes. Draft report. Kusatsu, Japan: International Lake Environment Committee.

Kira, T. (1997) "Survey of the state of world lakes", in Jørgensen, S.E. and S. Matsui, eds., *Guidelines of lake management, Volume 8: the world's lakes in crisis.* Kusatsu, Japan: International Lake Environment Committee and United Nations Environment Programme. Chapter 7.

Kurata, A. (1994) *Data book of world lake environments: a survey of the state of world lakes.* 5 Volumes. Kusatsu, Japan: International Lake Environment Committee, and Nairobi, Kenya: United Nations Environment Programme (UNEP).

Lundberg, J.G., M. Kottelat, G.R. Smith, M.L.J. Stiassny, and A.C. Gill. (2000) "So many fishes, so little time: an overview of recent ichthyological discovery in continental waters", *Annals of the Missouri Botanical Garden* 87: 26-62.

Loh, J. and M. Wackernagel (eds). (2004) The Living Planet Report 2004. Gland, Switzerland: WWF - World Wide Fund for Nature.

Master, L.L., S.R. Flack, and B.A. Stein (eds). (1998) *Rivers of Life: Critical Watersheds for Protecting Freshwater Biodiversity*, Arlington, Virginia, U.S.A.: The Nature Conservancy.

Moyle, P.B. and R.A. Leidy. (1992) "Loss of biodiversity in aquatic ecosystems: evidence from fish faunas", in P.L. Fiedler and S.K. Jain, eds., *Conservation Biology: The Theory and Practice of Nature Conservation, Preservation and Management.* New York, NY: Chapman and Hall, pp. 127-169.

Nilsson, C., C. Reidy, M. Dynesius, and C. Revenga. (2005) "Fragmentation and Flow Regulation of the World's Large River Systems", *Science* 308(5720): 405-408.

Postel, S. (1999) *Pillar of Sand: Can the Irrigation Miracle Last?* Washington, DC: WorldWatch Institute.

Revenga, C. and Y. Kura. (2003) Status and Trends of Biodiversity of Inland Water Ecosystems. Technical Series No. 11. Montreal, Canada: Secretariat of the Convention on Biological Diversity.

Revenga C, J. Brunner, N. Henninger, K. Kassem and R. Payne. (2000) *Pilot analysis of global ecosystems: freshwater ecosystems.* Washington, DC.: World Resources Institute, pp.83.

Reaka-Kudla, M.L. (1997) "The Global Biodiversity of Coral Reefs: A Comparison with Rain Forests", in M.L. Reaka-Kudla, D.E. Wilson, and E.O. Wilson, eds., *Biodiversity II: Understanding and Protecting Our Biological Resources*, Washington, DC: Joseph Henry Press, pp. 83–108.

Ricciardi, A. and J.B. Rasmussen. (1999) "Extinction Rates of North American Freshwater Fauna." *Conservation Biology* 13 (5): 1220–1222.

RHP (River Health Programme) (2003) State-of-Rivers Report: Free State Province Region River Systems. Pretoria, South Africa: Department of Water Affairs and Forestry.

Shiklomanov, I.A. (1997) "Comprehensive assessment of the freshwater resources of the world: assessment of water resource and water availability in the world." Stockholm, Sweden: World Meteorological Organization and Stockholm Environment Institute.

Smakhtin, V., Revenga, C. and Döll, P. (2004) Taking into account environmental water requirements in global-scale water resources assessments. Comprehensive Assessment Research Report 2. Colombo, Sri Lanka: Comprehensive Assessment Secretariat.

Sverdrup-Jensen, S. (2002) Fisheries in the Lower Mekong Basin: Status and Perspective. MRC Technical Paper No.6. Phnom Penh, Cambodia: Mekong River Commission.

UNESCO (United Nations Educational, Scientific, and Cultural Organization). (2000) *Water Related Vision for the Aral Sea Basin for the Year 2025*, Paris, France: UNESCO, pp. 237.

van Dijk, P.P., B.L. Stuart, and A.G.J. Rhodin. (2000) Asian Turtle Trade: Proceedings of a Workshop on Conservation and Trade of Freshwater Turtles and Tortoises in Asia Chelonian Research Monographs, No. 2, Chelonian Research Foundation in association with WCS, TRAFFIC, WWF, Kadoorie Farm and Botanic Gardens and the US Fish and Wildlife Service. Lunenburg, Massachusetts, USA: Chelonian Research Foundation, pp.164.

WHO (World Health Organization). (2004) *Water, Sanitation and Health: Facts and Figures 2004* available from http://www.who.int/water_sanitation_health/publications/facts2004/en/index.html

World Commission on Dams (WCD). (2000) *Dams and development: A new framework for decision-making*. London, U.K.: Earthscan Publishing Ltd.

WRI, IUCN, Ramsar and IWMI. (2003) *Watersheds of the World CD*, Washington D.C.: World Resources Institute.

WRI (World Resources Institute) in collaboration with United Nations Development Programme, United Nations Environment Programme, and the World Bank. (2000) "World Resources 2000-2001: People and Ecosystems: the Fraying Web of Life", Washington DC: World Resources Institute.

Managing Wetland Ecosystems – Balancing the water needs of ecosystems with those for people and agriculture

Max Finlayson

International Water Management Institute, Sri Lanka

Introduction

The importance of wetland ecosystems has been increasingly recognized as the ecosystem services that they provide are seen as essential for supporting the well-being of many people (Finlayson et al. 2005). This importance has been driven in part by greater realization that many wetland ecosystems have been degraded or lost along with the services they provided. It has also been driven by the increasing realization that this loss may be irreversible and may continue with dire consequences for biodiversity and people. At the same time, the debate about wetlands has been confused by differences in definitions and constrained by the extent and quality of basic information about the wetlands, their species and the services they provide (Finlayson & van der Valk, 1995; Finlayson et al., 1999).

The definition of wetlands varies greatly and this has caused confusion - the Ramsar Convention on Wetlands has adopted a broad definition that includes inland, coastal and near-shore marine ecosystems as diverse as rivers, lakes, swamps, marshes, estuaries and reefs, as well as human-made reservoirs and rice fields (paddy). Within the context of this paper, the discussion about wetlands is purposefully general and taken to include inland and some coastal aquatic ecosystems that provide food products, through agriculture and fishery, and/or fresh water.

Food and fresh water are among the many valuable ecosystem services provided by wetlands – the importance and status of these services has recently been assessed through the Millennium Ecosystem Assessment. The outcomes of the Assessment were especially poignant for wetlands as it was concluded that the degradation and loss of wetlands was more rapid than that of other ecosystems, and the status of freshwater and coastal wetland species was deteriorating faster than those of other ecosystems. Further, agricultural development has historically been the principal cause of the loss of inland aquatic systems worldwide and the construction of dams and other structures along rivers has resulted in fragmentation and flow regulation of almost 60% of the large river systems in the world (Finlayson & D'Cruz, 2005; Finlayson et al., 2005; Revenga & Kura, 2003).

Given that wetlands are also, *pro rata* in relation to their areal extent, amongst the most species diverse and productive ecosystems, the challenges facing wetland managers are explored. This is done from the context of increasing pressure for food production and competition for environmental allocations of water to maintain the already severely degraded ecological character of remaining wetlands.

Global Assessment of Wetlands

A few of the key messages developed by the Millennium Ecosystem Assessment for the continued provision of food and fresh water by wetlands (Millennium Ecosystem Assessment, 2005; Finlayson et al., 2005) are outlined before considering the challenges facing humankind in striking a balance between investing in water for food, ecosystems and livelihoods. The latter draws on a discussion paper prepared for the Comprehensive Assessment of Water Management in Agriculture (Molden & Fraiture, 2004).

The following points are taken from the General Synthesis of the Millennium Ecosystem Assessment (2005):

i) "Over the past 50 years, humans have changed ecosystems more rapidly and extensively than in any comparable period of time in human history, largely to meet rapidly growing demands for food, fresh water, timber, fiber and fuel."

ii) "The changes that have been made to ecosystems have contributed to substantial net gains in human well-being and economic development, but have been achieved at growing costs in the form of the degradation of many ecosystem services, ... and the exacerbation of poverty for some groups of people."

iii) "The degradation of ecosystem services could grow significantly worse during the first half of this century and is a barrier to achieving the Millennium Development Goals."

Changes in land cover and land use have resulted in major transformations in ecosystems and landscapes with substantial gains in human well-being from the production of food, but there are major concerns whether this is sustainable given the loss of other ecosystem services (Foley et al., 2005). The loss and degradation of wetlands has been driven by declines in the area and water quality along with increased invasion by alien species. The construction of dams and other structures along rivers has moderately or strongly affected flows in 60% of the large river systems in the world (Revenga et al., 2000) while water removal for human uses has reduced the flow of several major rivers, including the Nile, Yellow, and Colorado Rivers, to the extent that they do not always flow to the sea. As water flows have declined, so have sediment flows, which are the source of nutrients important for the maintenance of estuaries. Worldwide, sediment delivery to estuaries has declined by roughly 30% (Vörösmartry et al., 2003).

Management Challenges for Wetland Ecosystems and Human Well-Being

Within the scenario of declining wetland ecosystems, Molden & Fraiture (2004) consider the consequences of increased demand on ecosystem services as the world population increases by an estimated 2 billion more people by 2025. How much more water do we need to feed more people? Where does it come from and what are the environmental consequences? More food will be necessary and more food translates to more water being required for food production from both agriculture and aquaculture. Meeting the food and livelihood needs of a growing population will require some difficult choices about how water for agriculture (and to some extent aquaculture) is managed in the next few decades. At the same time there is increasing demand for water in rivers to be managed for environmental outcomes - how do we manage water for food *and* the environment? Can this be done in unison, or will it precipitate yet further competition? Will the allocation of more water for food production mean less for the environment and the further loss of other ecosystem services?

In simple terms these choices are often presented as choices between people and nature! Rijsberman & Molden (2001) state "It seems likely that the main competition for water over the next century will be between agriculture and the environment." Inland and coastal wetlands require an allocation of water (environmental flow) to ensure maintenance of ecosystem structure, functions and services. Pressure for such allocations may be increasing in some sectors, but they are still a low priority in water resource management.

When considering the scenario raised by Rijsberman & Molden (2001), a key message to promote is that wetlands provide many non-marketed and marketed benefits to people, and the total economic value of unconverted wetlands can be greater than that of converted wetlands. Finlayson et al. (2005) summarize examples where the economic value of intact wetlands, e.g. for fish production and storm protection, exceeds that of converted or otherwise altered wetlands. This does not mean that the conversion of wetlands is never economically justified, but it illustrates the fact that many of the economic and social benefits of wetlands have not been taken into account by decision-makers. Trade-offs between services have been made regularly, but often not consciously or transparently (Finlayson & D'Cruz, 2005).

Agricultural Demand for Water and Pressures on Wetlands

The push for greater agricultural production in recent decades has resulted in much land being converted from natural systems to cultivated systems (areas where at least 30% of the landscape is in croplands, shifting cultivation, confined livestock production, or freshwater aquaculture). Cultivated systems now cover one quarter of Earth's terrestrial surface.

In order to provide water to cultivated systems, natural water regimes have also been altered. In response to increasing demands for water, a large number of reservoirs has been constructed; the number of large dams in the world has increased from 5000 in 1950, to more than 45,000 at present, and store water for 30-40% of irrigated land and generate 19% of global electricity supplies (World Commission on Dams, 2000). Regulation of the water regime to this extent has caused significant degradation of wetlands/rivers, both inland and coastal (Finlayson et al., 2005). Dams have resulted in fragmentation and modification of aquatic habitats, transforming lotic ecosystems into semi/lentic ecosystems, altering the flow of matter and energy, and establishing barriers to migratory species movement. A global assessment of 227 major river basins showed 37% were strongly affected by fragmentation/altered flows, 23% moderately, and 40% unaffected (Revenga et al., 2000).

In many instances the trend for further land conversion and water regulation and/or increased food production continues, as shown by the trend for irrigated areas in developing countries and globally. Molden & Fraiture (2004) summarized projected increases in water withdrawals for irrigation for 2025 as ranging from 4-24% with the lower value being due to optimistic projections about increases in rain-fed areas and an assumption that increased food trade will play a major role in overcoming demands for more food globally. Producing more food also means using more water, and on past trends, continued environmental degradation of wetlands through further conversion and/or water regulation! Although irrigation development has alleviated poverty and improved human well-being it has had many negative environmental effects with the capacity of wetlands to provide the full range of ecosystem services having been drastically degraded (Gailbraith et al., 2005). The environmental trade-offs associated with irrigation have been increasingly articulated with Lemly et al. (2000) stating that "The conflict between irrigated agriculture and wildlife conservation has reached a critical point at a global scale".

The amount of water used for producing a range of food products has been calculated by various authors and summarized in Table 1. This shows some variability in water needs for food production, according to the type of food being produced as well as geographic differences. Taking these differences into account, each person is responsible for converting 2000 to 5000 litres of liquid water to vapour each day, compared to much lower amounts used for drinking -2 to 5 litres, and other household tasks, 50 to 200 litres. These figures illustrate the importance of food production in the overall water cycle. Based on these figures, it is clear that as consumption patterns continue to change, with increased demand for grain-fed meat products, these changes will further intensify water use and the demand for water.

Over the next 50 years, demand for food crops is projected to grow and food production to intensify (Wood & Ehui, 2005). Demand for water will therefore inevitably increase, with water withdrawals in developing countries increasing significantly and those in industrial countries declining (Vörösmartry et al., 2005). Substantial increases in the efficiency of water use and management may lessen the future demand for water, although achieving efficiencies may be made more complex by the consequences of global climate change. As an example, a scenario of vast and complex changes with great geographic variability in world freshwater resources may exacerbate the existing scenario where global freshwater use exceeds long-term accessible supplies, and irrigation withdrawals exceed supply rates and are therefore unsustainable.

Table 1: The amount of water (litres evaporated per kilogram) used to produce food (adapted from Molden & Fraiture, 2004).

	USA	France	China	India	Japan	World
Wheat	1 390	660	1 280	2 560	1 350	1 790
Rice	1 920	1 270	1 370	3 700	1 350	2 380
Maize	670	610	1 190	4 350		1 390
Beef*	10 060	7 740	12 600	14 379	9 540	9 680
Pork	3 370	1 940	2 520	7560	4 080	3 690

* figures are for grain-fed beef

Wetlands – Extent and Change

Estimates of the global extent of wetlands differ significantly and are highly dependent on the definition of wetlands used and on the methods for delineating wetlands. The *Global Review of Wetland Resources and Priorities for Wetland Inventory* estimated wetlands extent from national inventories as approximately 1,280 million hectares, which is considerably higher than previous estimates (Finlayson et al., 1999). This included inland and coastal wetlands (including lakes, rivers and marshes), near-shore marine areas (to a depth of 6 meters below low tide) and human-made wetlands such as reservoirs and rice paddies. Nevertheless, this figure is considered an underestimate, especially for the Neotropics and for certain wetland types (such as intermittently flooded inland wetlands, peatlands and artificial wetlands) where data were incomplete or not readily accessible. The data collated by Finlayson et al. (1999) suggest that the largest area of wetlands is in the Neotropics (32%), with large areas also in Europe and North America (Table 2). But figures provided by Lehner & Döll (2004) suggest that Asia may contain a greater and Europe a lesser area of wetlands (Table 2). The map showing wetland distribution produced by Lehner & Döll (2004) is presented in Figure 1.

Table 2: Estimates of global wetland area (million hectares) from the Global Review of Wetland Resources (Finlayson et al. 1999) and the Global Lakes and Wetlands Database (Lehner & Doll 2004).

Region	Global Review of Wetland Resources	Global Lakes and Wetlands Database
Africa	121-125	131
Asia	204	286
Europe	258	26
Neotropics	415	159
North America	242	287
Oceania	36	28
Total	**~1,280**	**917**

Information on peatlands, lakes, dams, major rivers, and rice fields is available, but is variable or lacking for many other types of wetland. Peatlands cover approximately 400 million hectares, with the vast majority in Canada (37%) and Russia (30%). There are several published inventories of rivers, but again there is considerable variability between estimates due to the method and definitions used. The area of rice fields has been estimated at about 130 million hectares, of which almost 90% is cultivated in Asia. Information on other human-made wetlands is variable and even lacking for many types. Information on the estimated 5-15 million lakes across the globe is also highly variable and dispersed. Many large lakes are found in Russia and North America, especially Canada. Tectonic belts, such as the Rift Valley in East Africa and the Lake Baikal region in Siberia, are the sites of some of the largest and most "ancient" lakes. Lakes have been mapped reasonably well although issues of scale also occur, with smaller lakes being more difficult to map. Some of the largest lakes are saline with the largest by far being the Caspian Sea (422,000 square kilometers).

Figure 1: Distribution of wetlands as shown by the Global Lakes and Wetlands Database compiled by Lehner and Doll (2004).

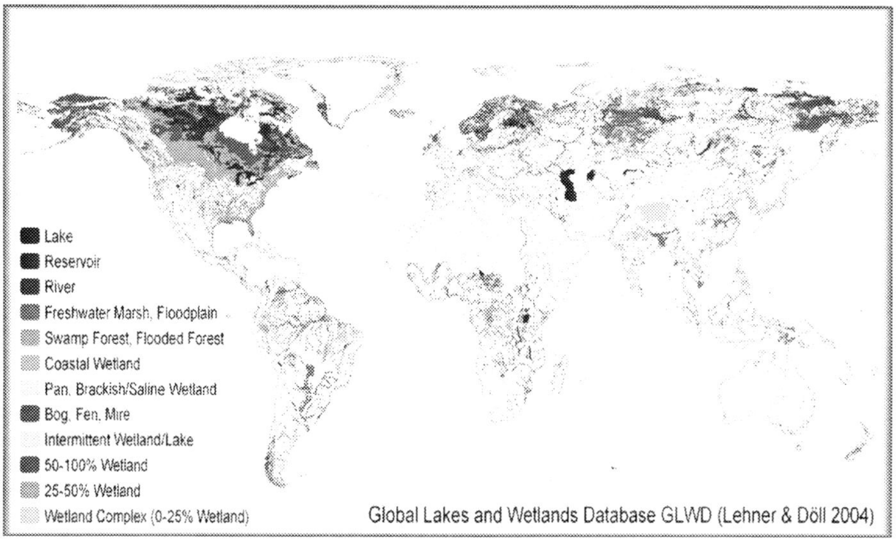

Information on the distribution of coastal wetlands such as estuaries and mangrove forests has been compiled (Agardy & Alder, 2005). The diversity of coastal habitat types and biological communities is significant, and the linkages between habitats are extremely strong, as is the extent of inter-connectivity with terrestrial systems and human settlements and infrastructure. Worldwide, there are some 1,200 major estuaries (those with a discharge of 10 cubic meters per second), with a total area of approximately 50 million hectares. Mangrove forests are found in both tropical and subtropical areas with global cover estimated at 16-18 million hectares and with the majority found in Asia.

The extent of wetland loss and degradation is widely recognized, but not often supported by quantitative data. More than 50% of specific types of wetlands in parts of North America, Europe, Australia, and New Zealand were lost during the twentieth century, but extrapolation of this estimate to wider geographic areas or to other wetland types is fraught with inaccuracy and can only be seen as speculation (Finlayson & D'Cruz, 2005). There is insufficient information available on the extent of all wetland types – such as inland wetlands that are seasonally or intermittently flooded, and some coastal wetlands – to document the extent of wetland loss globally. Much of the loss of wetlands has occurred in the northern temperate zone during the first half of the twentieth century, whereas many tropical and sub-tropical wetlands, such as swamp forests, were lost or degraded over the second half of the century (Finlayson & D'Cruz, 2005). As absolute measures of the condition of wetlands are hard to develop given the lack of baseline information, proxy indicators, such as the degree of fragmentation of rivers, have been used to infer the likely condition of at least some wetlands.

Although limited in area when compared with marine and terrestrial ecosystems many freshwater wetlands are relatively species-rich and support a disproportionately large number of species of certain faunal groups (McAllister et al., 1997). There is also evidence of a rapid and continuing widespread decline in many populations of wetland-dependent species, including mollusks, amphibians, fish, waterbirds, and some mammals (Finlayson & D'Cruz, 2005). An overall index of the trend in vertebrate species populations has also been developed showing a continuous and rapid decline in freshwater vertebrate populations since 1970 – a markedly more drastic decline than that for terrestrial or marine species (Loh & Wackernagel 2004). This information has been presented in the Living Planet Index and shows that freshwater populations have declined consistently and at a faster rate than the other species groups assessed, with an average decline of 50% between 1970 and 2000 (Figure 2). The index has a bias in the available data toward North American and European birds, and fish species other than commercial species are underrepresented. Even in the case of more poorly-known wetland fauna, such as invertebrates, assessments show that species in these groups are significantly threatened with extinction. A summary of information on the status of wetland species is shown in Table 3.

Figure 2: Trends in freshwater, marine and terrestrial Living Planet Index, 1970-2000 (from Loh & Wackernagel, 2004).

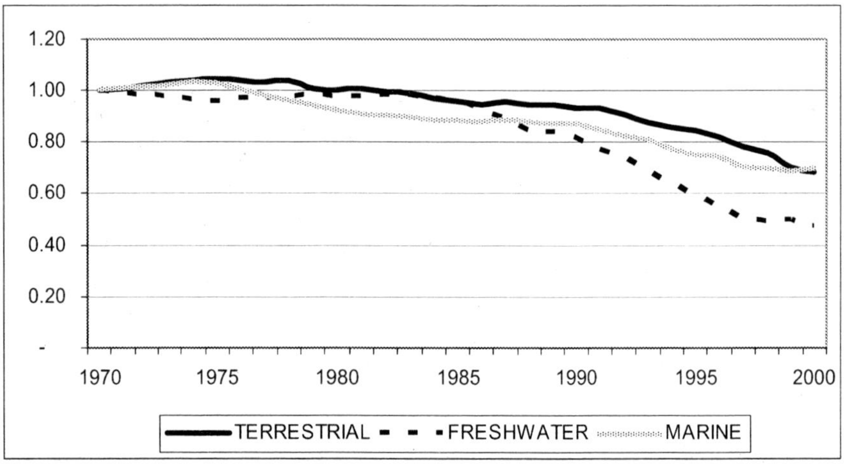

Table 3: Status and trends of major groups of wetland dependent species (adapted from Finlayson et al., 2005).

Species group	Status and Trends
Waterbirds	41% of the 1,138 biogeographic populations (of 868 waterbird species) of known trend are in decline. Of the 964 bird species that are predominantly wetland-dependent, 203 (21%) are extinct or globally threatened, with higher percentages of species dependent on coastal systems being globally threatened than are those dependent only on inland wetlands. The status of globally-threatened birds dependent on freshwater wetlands and even more so, that of coastal seabirds has deteriorated faster since 1988 than the status of birds dependent on other (terrestrial) ecosystems.
Mammals	Over one-third (37%) of the freshwater-dependent species that were assessed for the IUCN Red List are globally threatened; these include groups such as manatees, river dolphins, and porpoises, in which all species assessed are listed as threatened. Almost a quarter of all seals, sea lions, and walruses are listed in the IUCN Red List as threatened.
Freshwater Fish	Approximately 20% of the world's 10,000 described freshwater fish species have been listed as threatened, endangered extinct in the last few decades. In the countries for which assessments are most complete, an average of 17% of freshwater fish species are globally threatened. In addition a few well-documented cases show clearly this level of threat. The most widely known is the apparent disappearance of up to 123 haplochromine cichlids in Lake Victoria, although taxonomic questions remain a problem in accurately assessing this group of fish.
Amphibians	Nearly one-third (1,856 species) of the world's amphibian species are threatened with extinction, a large portion of which (964 species) are from freshwater, and especially flowing freshwater habitats. In addition, the population sizes of at least 43% of all amphibian species are declining, indicating that the number of threatened species can be expected to rise in the future.
Turtles	At least 50% of the 200 species of freshwater turtles have been assessed in the *IUCN Red List* as globally threatened and more than 75% of freshwater turtle species in Asia are listed in the as globally threatened, including 18 that are critically endangered, with one being extinct. All 6 species of marine turtles that have been assessed that use coastal wetlands for feeding and breeding are listed as threatened in the *IUCN Red List*.

Table 3 (continued): Status and trends of major groups of wetland dependent species (adapted from Finlayson et al., 2005).

Species group	Status and Trends
Crocodiles	Of the 23 species of crocodilians, which inhabit a range of wetlands including marshes, swamps, rivers, lagoons, and estuaries, 4 are critically endangered, 3 endangered, and 3 vulnerable.
Freshwater Crustacea and Mollusks	The IUCN Red List reports that some 275 species of freshwater crustacea and 420 mollusks are globally threatened, although no comprehensive global assessment has been made of all the species in these groups. In the United States, one of the few countries to comprehensively assess freshwater mollusks and crustaceans, 50% of known crayfish species and two-thirds of freshwater mollusks are at risk of extinction, and at least one in 10 freshwater mollusks are likely to have already gone extinct.
Dragonflies & Damselflies (Odonata)	A recent review of the global threat status of dragonflies and damselflies in 22 regions covering most of the world (except for parts of Asia) found relatively high levels of threat. In Australia 4 species are currently listed as globally threatened but 25 species are considered to be in critical condition, with another 30% species being data-deficient. In North America, 6% (25 species) are considered to be of conservation concern. In the Neotropics, 25 species are already considered globally threatened and a further 45 species are considered of high conservation priority, with many others data-deficient.

The primary indirect drivers of degradation and loss of wetlands have been population growth and increasing economic development often through increased food production and water regulation. The primary direct drivers include infrastructure development, land conversion, water withdrawal, pollution, over-harvesting and over-exploitation, and the introduction of invasive alien species. Summaries of the extent of the pressure exerted by these drivers are given by Finlayson & D'Cruz (2005) and Finlayson et al. (2005). Figure 3 is a pictorial representation of some of the direct drivers of change in inland and coastal wetlands. Invasive species, climate change, and land conversion to urban or sub-urban areas affect all components of the catchment and coastal zone, and are not represented pictorially.

Figure 3: Pictorial representation of the direct drivers of change in wetlands. Invasive species, climate change, and land conversion affect all components of the catchment and coastal zone, and are not represented pictorially (from Finlayson & D'Cruz 2005).

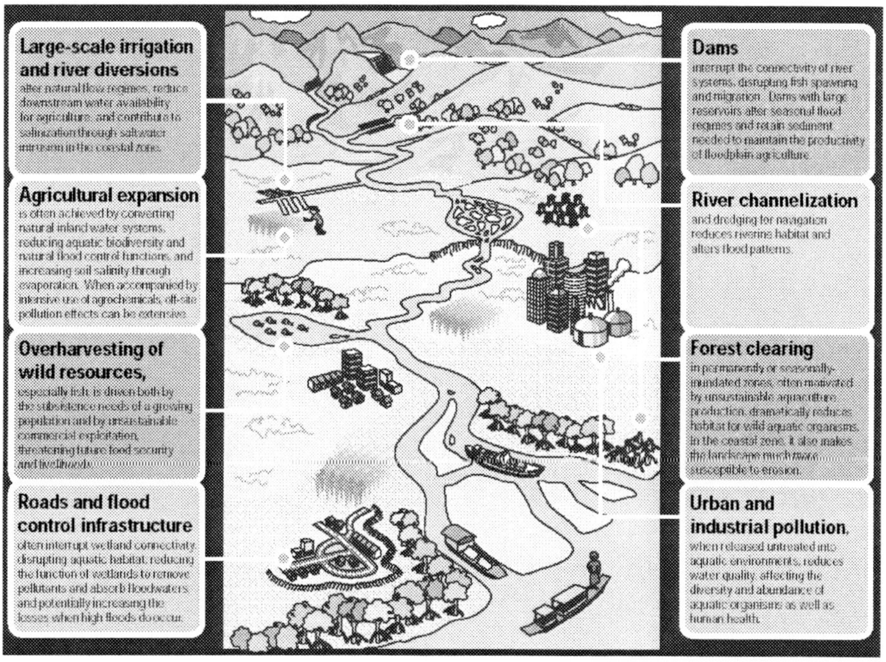

Clearing and drainage, often for agricultural expansion, and increased withdrawal of fresh water are the main reasons for the loss and degradation of inland wetlands such as swamps, marshes, rivers, and associated floodplain water bodies. By 1985, an estimated 56-65% of inland and coastal marshes (including small lakes and ponds) had been drained for intensive agriculture in Europe and North America, 27% in Asia, 6% in South America, and 2% in Africa. Agricultural systems and practices have exerted a wide range of mostly adverse impacts on inland and coastal wetlands globally. Both the extensive use of water for irrigation and excessive nutrient loading associated with the use of nitrogen and phosphorus in fertilizers have resulted in a decline in the delivery of services such as fresh water and some fish species. The introduction of invasive alien species is now considered to be a major cause of extinction of native freshwater species.

A pictorial summary of the impact of each of the main drivers on the biodiversity of wetlands over the past 50-100 years is shown in Figure 4. The intensity of impact is shown by colours with high impact meaning that over the last century the particular driver has significantly altered the biodiversity, and low impact indicating that it has had little influence on biodiversity. Arrows are used to indicate the trend in the drivers. Thus, for example, if a wetland had experienced a moderate impact of a particular driver in the past century (such as the impact of overexploitation in inland water systems), a horizontal arrow indicates that this moderate high impact is likely to continue. The information in Figure 4 is based on expert opinion consistent with and based on the analysis of drivers of change in the Millennium Ecosystem Assessment and summarized for wetlands in Finlayson et al. (2005). The global impacts and trends shown in Figure 4 may be different from those in specific regions.

Figure 4: Main direct drivers of change in wetland systems.
The cell color indicates the impact of each driver on biodiversity in wetlands over the past 50–100 years. The arrows indicate the trend in the driver. Horizontal arrows.

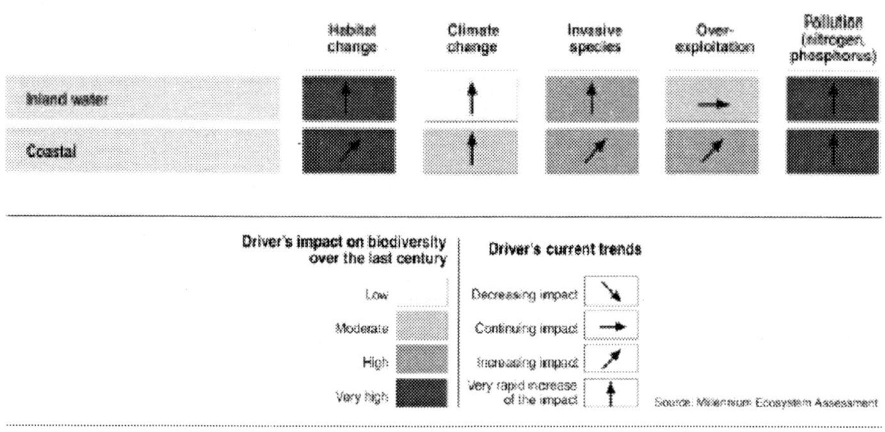

Management Responses

The messages raised above have most recently been highlighted in the Millennium Ecosystem Assessment, but many are not newly discovered – they have on the whole been raised in many previous reports and fora. One example is given by the analyses presented at the 1991 Mediterranean wetland forum held in Grado, Italy (Finlayson et al., 1992). Throughout this forum the issues of biodiversity decline due to agricultural expansion, regulation of water, and pollution, etc., were raised, as were the problems of inadequate and sectoral management responses. Further, Hollis (1992) articulated the need to address the indirect causes of wetland loss and degradation and to make the following statements:

- "This loss and degradation is rooted in social, economic and political processes."
- "There has to be an offensive on the social, economic and political causes of wetland loss and degradation."

The outcomes of the Millennium Ecosystem Assessment lend support to these statements and it is as true today as in the early 1990s that wetland degradation can rarely be reversed without actions that address one or more of the indirect drivers of change. Thus, management responses in support of the wise use of wetlands need to encompass or lend support to actions that address the following:

- population change, including growth and migration;
- change in economic activity, including economic growth, disparities in wealth, and trade patterns;
- economic policies and practices that distort markets or encourage ineffective land and water management;
- sociopolitical factors, including factors ranging from the presence of social conflict to public participation in decision-making;
- cultural factors that enhance or inhibit wise use of wetlands in a contemporary era;
- technological change that can enhance production and reduce the pressure on remaining wetlands; and
- changing knowledge bases and incorporation of relevant information (from multiple sources, including traditional societies) into management.

Whilst specific technical responses are required at multiple scales and locations, it is contended that unless the indirect drivers are addressed the chances of reversing at a global scale the past trends that have resulted in extensive loss and degradation of wetlands are slim.

In the Millennium Ecosystem Assessment wetland synthesis Finlayson et al. (2005) recognize the social issues that underpin better wetland management and highlight the role of stakeholder participation in addressing these issues. They conclude with the following statements about management responses: "Stakeholder participation (also) contributes to the decision-making process because it allows for a better understanding of impacts and vulnerability, the distribution of costs and benefits associated with trade-offs, and the identification of a broader range of response options that are available in a specific context. And stakeholder involvement and transparency of decision-making can increase accountability and reduce corruption."

The Future – Concluding Comments

In concluding, a few key statements are presented as a basis for further analysis and management responses. These statements build on the points raised above and by Molden & Fraiture (2004). They introduce some further issues that need to be considered if we are to make better use of our wetlands for the long-term:

- In the past there has been too much complacency about water and food issues – this needs to be reversed through effective social and managerial responses that address both indirect and direct drivers of change;
- Increased pressure from increased population and consumption of food will translate to increased pressure on wetlands and further loss of wetland species and services currently available to many people;
- Business as usual will result in more competition (or conflict) between food production and nature and result in more environmental degradation, and possible persistence of poverty and food insecurity;

- Improving irrigation efficiency may make gains for food production, but may not enhance the surrounding environment as in many basins little water is being wasted – leakages from irrigation systems currently can be captured, but with likely adverse affects on groundwater recharges or flows that currently occur back to wetlands;
- Improving productivity through improved crop yields through irrigation may result in increased use of fertilizers and agricultural chemicals and increased pollution and eutrophication of many types of wetlands;
- Increases in water productivity and upgrading of rain-fed food production hold a lot of promise for increasing food production and could occur given better governance, policies and institutions.

As alluded to in the point immediately above, major changes in agricultural and environmental management may not be possible without successful outcomes from increased investment in institutions and policies. It is recommended that these support the integration of ecosystem management goals within other business/economic and governance sectors and within broader development planning frameworks that encompass changes in existing economic practices.

Reference has already been made to the need for increased transparency and accountability of government sector performance and the elimination of subsidies that promote excessive use of ecosystem services. Perhaps more socially controversial are calls for greater use of economic instruments and market-based approaches in the management of ecosystem services – the social and cultural implications of such change need investigating. It should not be assumed that market-based approaches are a panacea for environmental management even though it is widely recorded that an absence of markets for many environmental services has contributed to past loss and degradation. The development of market mechanisms could in some instances support greater conservation of services provided by wetlands – in other instances they may not. Many services provided by wetlands are currently freely available – recognizing their worth through economic valuation, for example, need not mean that they should be incorporated into market mechanisms.

As a final comment, the following is proffered. Can we achieve sustainable development of wetlands – can we put our concepts into practice and develop wetland resources and conserve them at the same time? Is it possible to develop wetlands without further running down the resource and the very processes and interactions that support the wetland? Many wetland resources are already depleted and human populations and consumption are increasing. The Millennium Ecosystem Assessment has provided an analysis – what is the practice?

References

Agardy, T. and J. Alder. (2005) "Coastal Systems", in *Millennium Ecosystem Assessment, Conditions and Trends*. Washington, D.C., USA: Island Press, (in press).

Finlayson, C.M. and A.G. van der Valk (eds). (1995) *Classification and Inventory of the World's Wetlands (Advances in Vegetation Science 16)*, Dordrecht, The Netherlands: Kluwer Academic Press.

Finlayson, C.M. and R. D'Cruz. (2005) "Inland Water Systems" in *Millennium Ecosystem Assessment, Conditions and Trends*. Washington, D.C., USA: Island Press, (in press).

Finlayson, C.M., R. D'Cruz and N.C. Davidson. (2005) Ecosystem Services and Human Well-being: Wetlands and Water Synthesis, Washington D.C., USA: World Resources Institute.

Finlayson, C.M., T. Hollis and T. Davis (eds). (1992) Managing Mediterranean Wetlands and Their Birds. Proceedings of an IWRB International Symposium, Grado, Italy, February 1991. IWRB Special Publication No. 20, Slimbridge, UK. pp. 285.

Finlayson, C.M., N.C. Davidson, A.G. Spiers and N.J. Stevenson. (1999) "Global wetland inventory - status and priorities", *Marine and Freshwater Research* 50: 717–727.

Foley, J.A., R. DeFries, G.P. Asner, C. Barford, G. Bonan, S.R. Carpenter, F.S. Chapin, M.T. Coe, G.C. Daily, H.K. Gibbs, J.H. Helkowski, T. Holloway, E.A. Howard, C.J. Kucharik, C. Monfreda, J.A. Patz, I.C. Prentice, N. Ramankutty, and P.K. Snyder. (2005) "Global consequences of land use", *Science* 309: 570-574.

Gailbraith, H., P. Amerasinghe, and A. Huber-Lee. (2005) The effects of agricultural irrigation on wetland ecosystems in developing countries: a literature review. Comprehensive Assessment Discussion Paper 1. International Water Management Institute, Colombo, Sri Lanka.

Hollis, G.E. (1992) "The causes of wetland loss and degradation in the Mediterranean", in C.M. Finlayson, G.E. Hollis and Davis, T.J., eds., *Managing Mediterranean Wetlands and Their birds, being: Wetlands and Water Synthesis*. World Resources Institute, Washington, D.C. USA, IWRB Special Publication No. 20, Slimbridge, UK, pp. 83–90.

Lehner, B. and P. Döll. (2004) "Development and validation of a global database of lakes, reservoirs and wetlands", *Journal of Hydrology* 296: 1–22.

Lemly, A.D., R.T. Kingsford, and J.R. Thompson. (2000) "Irrigated agriculture and wildlife conservation: conflict on a global scale", *Environmental Management* 25: 485–512.

Loh, J, and M. Wackernagel (eds). (2004) Living Planet Report 2004. WWF-World Wide Fund for Nature, Gland, Switzerland, pp. 42.

McAllister, D.E., A.L. Hamilton, and B. Harvey. (1997) "Global freshwater biodiversity: striving for the integrity of freshwater ecosystems", *Sea Wind - Bulletin of Ocean Voice International* 11: 1–140.

Millennium Ecosystem Assessment (2005) Millennium Ecosystem Assessment Synthesis Report. Washington, D.C., USA: Island Press.

Molden, D. and C. de Fraiture. (2004) Investing in water for food, ecosystems and livelihoods. Comprehensive Assessment of Water Management in Agriculture, Colombo, Sri Lanka.

Revenga, C. and Y. Kura. (2003) Status and Trends of Biodiversity of Inland WaterEcosystems. Secretariat of the Convention on Biological Diversity, Montreal, Technical Series No. 11, pp. 120.

Revenga, C., J. Brunner, N. Henninger, K. Kassem and R. Payne. (2000) *Pilot Analysis of Global Ecosystems: Freshwater Systems.* Washington D.C.: World Resources Institute, pp. 83.

Rijsberman, F.R. and D.J. Molden. (2001) Balancing water uses: water for food and water for nature. Thematic background papers. International Conference on Freshwater, Bonn, Germany, December 2001, pp. 43-56.

Vörösmarty, C.J., C. Lévêque and C. Revenga. (2005) "Fresh Water", in *Millennium Ecosystem Assessment, Conditions and Trends.* Washington, D.C., USA: Island Press, (in press).

Vörösmartry, C.J., M. Meybeck, B. Fekete, K. Sharma, P. Green, and J.P.M. Syvitski. (2003) "Anthropogenic sediment retention: major global impact from registered river impoundments", *Global and Planetary Change* 39: 169-190.

Wood, S. and S. Ehui. (2005) "Food", in *Millennium Ecosystem Assessment, Conditions and Trends.* Washington, D.C., USA: Island Press, (in press).

World Commission on Dams (WCD). (2000) *Dams and Development: A new Framework for Decision-making. The Report of the World Commission on Dams*, London U.K.: Earthscan Publications Ltd.

Ecohydrology as a Concept and Management Tool

Maciej Zalewski

International Centre for Ecology, Polish Academy of Sciences, Poland

Introduction

The freshwater ecosystems are situated in depressions in the landscape. As a consequence, water quality has been, to a great extent, dependent on human population density and its whole range of activities. The anthropogenic impact on freshwater environments can be defined in two dimensions: chemical alteration of water quality through emission of pollutants and physical degradation of environments. While the emission of pollutants can be controlled by technologies, restoration of physically degraded environments where modification of hydrological and biogeochemical cycles has taken place due to e.g., deforestation, urbanisation, canalisation, requires a new, systemic approach.

Ecohydrology (EH) is a new paradigm developed over the lifetime of UNESCO IHP-V, 1996-2001 (Zalewski et al., 1997). It has been a new concept which suggests that the sustainable development of water resources has been dependent on the ability to maintain the ecosystem processes that have been established by evolution. These ecosystem processes regulate water and nutrient circulation and energy flow at the basin scale in two ways: biota are regulated by hydrology and hydrology is regulated by biota (Zalewski, 2000). The general question posed in the formulation of the Ecohydrology concept concerns how to regulate the biological processes of freshwater ecosystems using hydrology; and, vice versa, how to use biotic ecosystem properties as a tool in water management. Syntheses of this concept, describing the hypotheses, tenets and examples of applications of ecohydrology have been published by UNEP and UNESCO (Zalewski, 2002, Zalewski et al., 2004).

The Hypotheses of Ecohydrology

In the face of increasing pressure on freshwater resources, there remains an urgent need for new practical tools to achieve their sustainable management. Traditional approaches to water management do not consider ecosystem properties as potential management tools. During the genesis of ecohydrology, it was concluded that a new type of hypotheses should be used by scientists in order to better understand and address management problems in freshwater ecosystems. These hypotheses should lead to the integration of both biota and hydrology in management solutions. Such hypotheses should meet the two following fundamental conditions:

i) They should be related to the dynamics of the two entities (biota and hydrology) in way that would be impossible to answer without consideration of both of the two components (both ways E↔H). In other words, this question should enable the defining of relationships between hydrological and biological processes in order to obtain comprehensive empirical data at the same spatial and temporal scales.

ii) The results of the empirical analysis should test the whole range of processes (from a molecular to catchment scale), should enable their spatio-temporal integration and should be convertible to large-scale management measures in order to enable further testing of the hypotheses.

Taking into account the above conditions, the key hypotheses for ecohydrology have been defined based on an in-depth understanding of the interplay between biological and hydrological processes and the factors that regulate and shape them:

Hypothesis H1: "The regulation of hydrological parameters in an ecosystem or catchment can be manipulated in order to control biological processes".

Hypothesis H2: "The shaping of the biological structure of an ecosystem(s) in a catchment can be manipulated in order to regulate hydrological processes".

Hypothesis H3: "Both types of regulation integrated at a catchment scale and in a synergistic way can be used to achieve the sustainable development of freshwater resources. Sustainable management can be measured as the improvement of water quality and quantity (providing ecosystem services)".

It should be stressed that according to the ecohydrology concept, the overall goal defined in the above hypotheses is the sustainable management of water resources. This should be focused on the enhancement of ecosystem carrying capacity for ecosystem services and anthropogenic stress. Such an interdisciplinary, integrative approach provides the background to convert environmental threats into sustainable development (see Zalewski and Robarts, 2003).

The Principles of Ecohydrology

The concept of Ecohydrology is based on the following principles, providing framework, target (goal) and methodology for the concept (Zalewski et al., 1997; Zalewski, 2000):

1. **Integration of catchment and biota into a Platonian superorganism** for identification of threats and opportunities for reversing degradation of water and ecosystem resources. This covers such aspects as:

- Scale - the mesocycle of water circulation in a basin (terrestrial/aquatic ecosystem coupling) has been the template for quantification of ecological processes such as nutrient dynamics and energy flow;
- Dynamics - water and temperature has been a driving force for terrestrial and freshwater ecosystems;
- Hierarchy of factors - the abiotic processes are dominating (hydrology), however once they become stable and predictable the biotic interactions begin to manifest themselves (Zalewski and Naiman, 1985).

2. Understanding the Evolutionarily-Established Resistance and Resilience of Such Superorganisms to Stress

This aspect of Ecohydrology expresses the proactive approach for sustainable management of freshwater resources. It assumes that it is not enough simply to protect ecosystems by regulating or preventing human activities within them. Global changes that are associated with increases in population, energy, material and human aspirations are inevitable. It is therefore necessary to concomitantly increase the 'absorbing capacity' (resistance and resilience) of ecosystems against these human impacts.

3. The Use of Ecosystem Properties as Management Tools

The use of ecosystem properties as management tools has been termed Ecological Engineering (Mitsh, 1993; Jørgensen, 1996). Ecohydrology is a systems approach implemented within IWRM. The third principle is related to ecological engineering and defines how to enhance resilience and resistance of ecosystems in river basins based on ecohydrology tenets.

The three key tenets of ecohydrology were formulated as a result of several years of cooperation by an interdisciplinary team of scientists involved in the UNESCO International Hydrological Programme:

i) ***Dual Regulation*** of hydrology by shaping biota and, vice versa, regulation of biota (eg., elimination of toxic algal blooms) by altering hydrology (eg., Zalewski et al., 1990; Zalewski et al., 2001) – Figure 1.

ii) ***Integration*** at the basin scale various types of regulations (E↔H) should be integrated towards achieving synergy to stabilize and improve the quality of freshwater resources at a basin scale (Zalewski, 2000).

iii) ***Harmonization*** of ecohydrological measures with necessary hydrotechnical solutions (e.g., dams, irrigation systems, sewage treatment plants, levees at urbanized areas etc.) - Hellegers and Witte, 2002; Timchenko and Oksiyuk, 2002).

Empirical testing of the three tenets of ecohydrology at different river basins is urgently needed toward their implementation for sustainable development.

Eutrophication is an example of the problem that can be addressed by application of ecohydrology. The eutrophication of inland water is a most complex consequence of various forms of human impact, which are synergistically amplified at the whole catchment "super organism" level. These synergistic impacts include the effects of agriculture, urbanization, recreation, point source pollution. Due to its complexity, the eutrophication process was considered as a primary case for formulation of ecohydrological principles and tenets. However, ecohydrology was formulated toward a universal solution. Ecohydrology provides a conceptual framework to encourage scientific research and innovation and leads to improved management through the wider use of ecological engineering. Ecological engineering provides a framework to solve problems in all types of ecosystems, based on tactical and operational level. Ecohydrology refers to river basins in particular, and expends the approach to a vision (enhancing carrying capacity of ecosystems – III principle of ecohydrology), strategy (I principle of ecohydrology) and a new element of ecological engineering based on dual regulation between hydrology and biota in a basin scale.

There is an increasing body of evidence confirming the validity of the application of ecohydrology to a broad range of issues related to water resources management. Among others, ecohydrology has been applied eg., for enhancement of biodiversity (Agostinho, et al., 2001), fish production (Timchenko and Oksiyuk, 2002), bioenergy, reduction of stable pollutants (eg. heavy metal and pesticides) (Gouder de Beauregard and Mahy, 2002), socio-economic feedbacks (Zalewski, 2002) in ecosystems.

1. Dual Regulation between Hydrology and Biotic Processes

The Use of Hydrology for Regulation of Biotic Interactions

The first of three tenets of the ecohydrological approach – regulation of a biological cascade by hydrological manipulation - is exemplified by a case study in Sulejow Reservoir (Figure 1).

Under conditions of eutrophication, fishes depend for food on the limnetic zone, but still reproduce in the littoral. As a result, in eutrophicating reservoirs where planktonic food is not limited, the reproductive success of cyprinids, percids and centrarchids depends mostly on spawning substrata, especially on the extent to which shore vegetation is flooded (Ploskey, 1985). Following flooding of the shoreline herbal vegetation, fry survival was high, large zooplankters were drastically reduced, planktonic algal biomass increased sharply, and water quality declined. Due to overcrowding, intra- and interspecific competition among fry was high, resulting in mass shoreline migration, 30% growth retardation, and low overwinter survival. The scarcity of large zooplankters reduced growth so much in pike-perch fry (Stizostedion lucioperca L.) that they were not big enough to eat even the slow growing perch (Perca fluviatilis L.) and roach fry in mid-July. Usually, at this time they would normally become piscivores. The consequent lack of one generation of such an easily over-exploitable species might reduce its population density seriously for many years (Reid and Momot, 1985).

So, in temperate lowland reservoirs the reproductive success of dominant fish species could be regulated by their access to the shoreline ecotone. This could be adjusted by the hydrological regulation of a dam. On the other hand, by enhancing piscivores to reduce the planktivorous fish populations (Hrbacek et al., 1961; Shapiro et al., 1975), the density of efficient large filtering zooplankton (eg., cladocerans) can be increased, and the quality of stagnant water improved.

Figure 1: The first tenet of ecohydrology: "**Dual Regulation**": The regulation of a cascade of biological processes by hydrological manipulation for improvement of water quality and optimization of fish yield (for explanation see Zalewski et al., 1990).

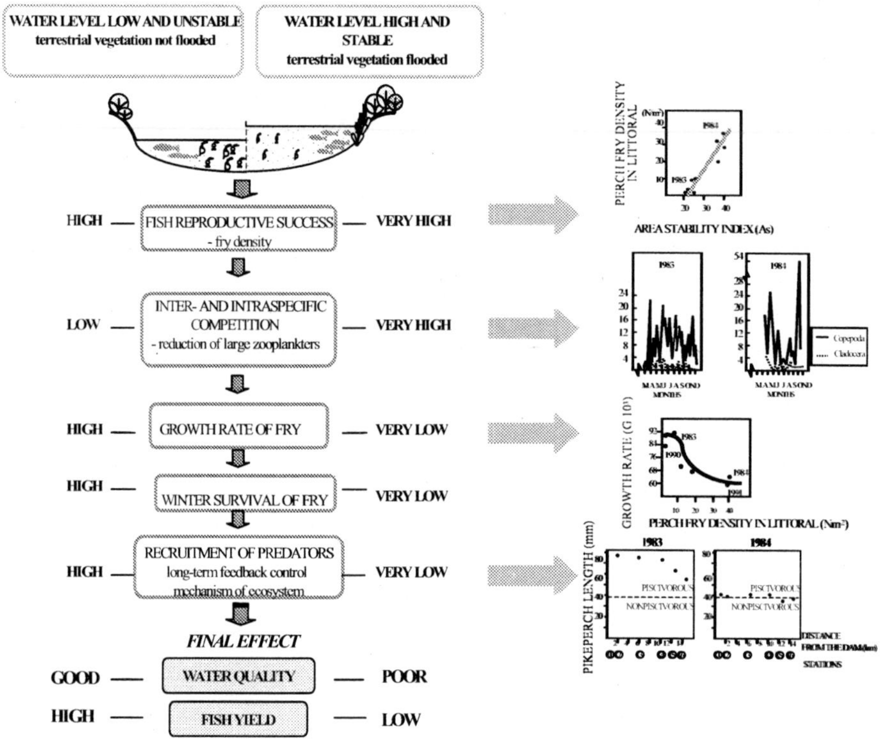

Use of Biological Structure Dynamics for Regulation of Hydrological Processes

River floodplains play a crucial role in the exchange of water masses and matter between the river and terrestrial ecosystems and in the functioning of the river ecosystem (Junk et al., 1989). If appropriately shaped and managed, floodplain vegetation can be an efficient ecohydrological tool for the regulation of floodplains hydraulics, river hydrology and optimisation of nutrient retention (Wagner-Łotkowska et al., 2004). This optimisation is based on both physical and chemical processes and biological retention of nutrients in biomass, leading to shortening of their spiraling loop.

Analysis of the variability of the chemical composition of waters of the major tributaries of a lowland reservoir in Poland (Sulejow) on the background of its hydrological pattern enabled the identification of episodes of the highest nutrients transport and thus reservoir supply, varying depending on flood characteristics (Wagner & Zalewski, 2000, Zalewski et al., 2000). If the river's water in these periods is directed into floodplains, selfpurification of the river increases (Wagner & Zalewski, 2000). Thus, the control of the river's discharge can reduce downstream nutrient transport and modify pattern and intensity of biological processes in the reservoir.

Vegetation cover on the floodplain influences water flow, matter sedimentation and also biomass production and nutrients assimilation in biomass. The latter varies by plant species, vegetation communities, timing of flooding and hydroperiod (Kiedrzynska, Wagner-Lotkowska, Zalewski, in press). Thus creating intermediate patches of diversified vegetation adapted to hydrological characteristics can enhance the ability of the system to retain nutrients. Research carried out at the experimental floodplain of the Pilica River in central Poland showed, that covering of 24% and 48% of a floodplain of about 27ha with willow patches, can increase it's capacity for phosphorus accumulation from 250,0 kg P y-1 to 340,6 kg P y-1 and 399,4 kg P y-1 respectively (Figure 2, Kiedrzynska, Wagner-Lotkowska, Zalewski, in press).

Figure 2: Calculations of phosphorus accumulation by plant communities and enhancement of absorbing capacity of a floodplain for nutrients by increasing willow contribution, up to 24% and 48% (Kiedrzyńska E., Wagner-Łotkowska I., Zalewski M., in press).

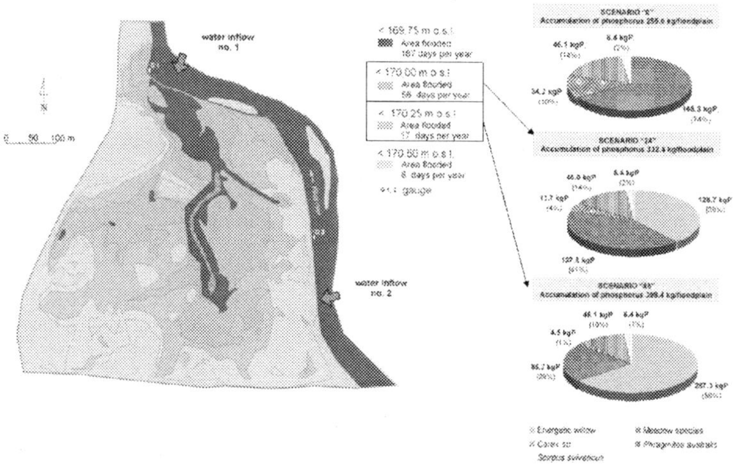

Using autochthonous vegetation as the major component maintains biodiversity in river corridors, improving their functioning and efficiency of selfpurification. Introduction of autochthonous species of willows can additionally provide some economic benefits, and be a potential source of biomass for the provision of bioenergy. In order to increase the efficiency of the system, the managed vegetation should be seasonally removed. It has been suggested to do it every 3 years in the case of willows (Wagner & Zalewski, 2000). This is because willows maintain the highest growth rate and effectiveness of phosphorus uptake within this period (Zalewski, 2002). Removing vegetation after the growth season prevents the release of nutrients back into the water in autumn. Willow could be utilized as an alternative crop especially for these floodplains which have undergone intense agricultural cultivation, causing negative effects on water quality.

2. Integration of Hydrological and Ecological Regulatory Measures in Basin Scale

The integrative application of all three principles at river basin level has been exemplified by Figure 2, illustrating the reduction of eutrophication in a temperate reservoir. Different ecological processes in the river basin have been integrated in order to reduce phosphorus input from the terrestrial part of the catchment freshwater and reduction of the dynamic pool of phosphorus, i.e., these chemical forms that can be easily assimilated by organisms and incorporated in into the biological structure of the ecosystem.

Starting from the top of the catchment, the first stage has to be the enhancement of nutrient retention within the catchment by reforestation, creation of ecotone buffering zones and optimization of agricultural practices. The buffering zones at the land-water interface reduce the rate of groundwater flux due to evapotranspiration along the river valley gradient. Nutrient transformation into plant biomass in ecotone zones may further reduce the supply into the river. The wetlands in the river valley from the buffering zone reduce the mineral sediments, organic matter and nutrient load transported by the river during floods periods through sedimentation. Also, in some artificial wetlands, nitrogen load can be reduced significantly by regulation of the water level to stimulate denitrification through anaerobic processes. In shaded rivers with high nutrient loads, it is possible to amplify the self-purification capacity by increasing light access and maintaining filtering function - creating intermediate complexity of ecotones. If, despite all of the above measures combined with necessary sewage treatment plants, the nutrient concentrations in the reservoir are too high and potentially might be converted into toxic algal blooms, there exist numerous methods to reduce recirculation of nutrients in the reservoir by blocking it in the biomass of macrophytes, and translocation between trophic levels (e.g. biomanipulation).

Since the properties of a large scale system cannot be predicted from properties of its component elements, such a complex strategy for restoring and controlling nutrients in the catchment landscape and freshwater ecosystem should be assessed continuously at every stage of implementation (Holling et al., 1994) and adjusted to maximize the potential for synergistic effects.

Figure 3: The second tenet of ecohydrology – synergistic "**Integration**" of various ecohydrological measures for system regulation in a catchment.

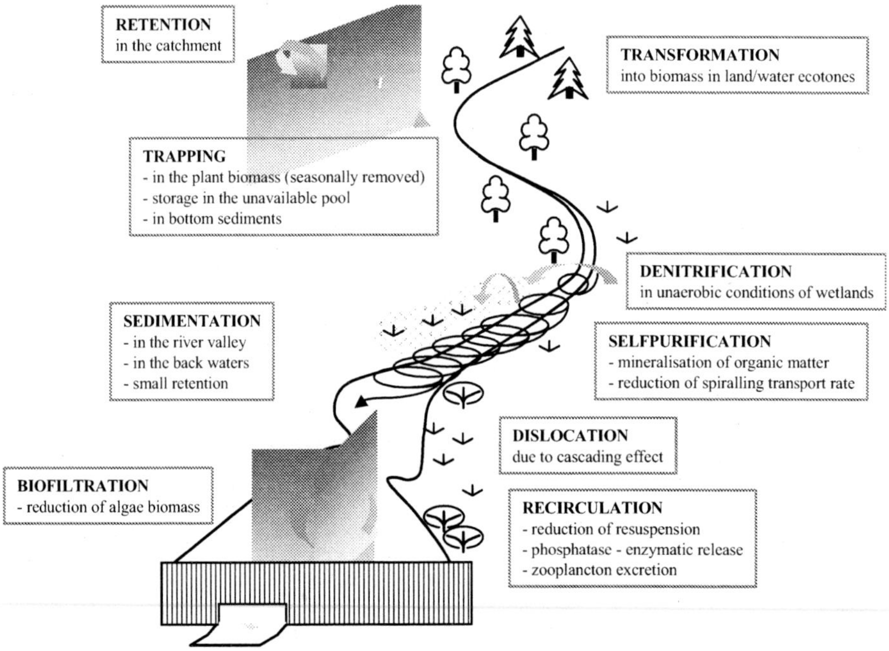

3. Harmonization of Ecohydrological Measures with Necessary Hydrotechnical Solutions: Sewage Treatment Plants Integrated with Wetlands

Traditional sewage treatment plants in a small town usually do not possess a sophisticated tertiary chemical treatment stage. This is due to the high construction costs of such systems, which local communities cannot afford. The most widely used treatment plant models, therefore, reduce BOD and some nutrients, but still negatively influence water quality, reducing the benefits of rivers and reservoirs and their recreational values. Extending the sewage treatment by constructing a wetland results in the more efficient reduction of pollutant loads and generates additional societal benefits. Improvement of water quality increases the appeal of water resources for tourism, which contributes to the inflow of capital to a region (Figure 4).

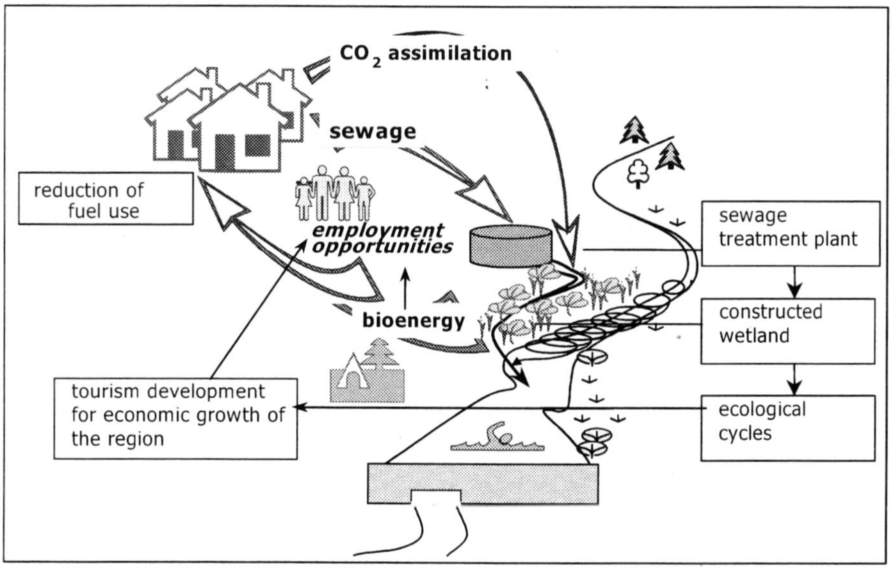

Figure 4: The third tenet of ecohydrology – **"Harmonization"** of hydrotechnical structure with ecosystem and local populations. The development of the ecohydrology concept for improving water quality, ecosystem services and the creation of positive socioeconomic feedbacks.

There is scope for further benefits to be introduced into the harmonized solutions. For example, the establishment of multispecies willow plantations can bring additional benefits beyond the improvement in water quality. Local species can tolerate the resulting high groundwater level, maintain river valley landscape biodiversity and provide an alternative source of energy (bioenergy) that can help to reduce CO_2 emissions from burning fossil fuels. The resultant ash can be used to fertilize forest plantations. Thus, pollutants are converted into bioenergy. Producing bioenergy and timber also generates new employment opportunities and revenue flows while reducing capital outflows for fossil fuel use. Bioenergy can be used for conversion of non-degradable plastic wastes by low energy technology in to paraffin's linking water and waste management. The use of ecological knowledge, therefore, results not only in a good quality environment but also can help to elevate the economic status and level of sustainable development in local communities.

Such an implementation case as a UNESCO/UNEP demo site has been recently under development at the town of Przedborz on the Pilica River, a western tributary of the Vistula River, above Sulejow Reservoir in Poland (Figure2).

Conclusion

The degradation of freshwater ecosystems has been of a two-dimensional character, comprising both chemical pollution and the physical disruption of water and biogeochemical cycles established in the landscape by evolution. Both cause destruction of the biotic structure of the catchments, of ecosystem services and of the freshwater ecosystems themselves, and lead to water resources decline.

Pollution can be significantly reduced or eliminated by technological progress. However, degradation of natural processes on the catchment scale creates much more complex problems. Natural processes that can be disrupted by pressures on ecosystems include cycles for water circulation, nutrient and energy flows.

The 21st century will become an era of integrative science because understanding the complexity of our world is a key to achieving sustainable development. This is especially urgent in ecology and environmental sciences for two reasons. First, there is an urgent need for sound solutions to address declining ecosystem services and biodiversity at the global scale. The second is that further scientific progress can be made by testing existing concepts and "know-how", by implementing concepts and methods integrated at basin- and landscape scales.

Acknowledgments

To Caroline King, for thoughtful comments and suggestions which to great extent helped to clarify many aspects of ecohydrology. To my co-workers, especially Iwona Wagner, Piotr Frankiewicz, Boguslawa Brewinska-Zaras and Edyta Kiedrzynska, with whom field work and discussions at different stages of development and formulation of ecohydrology concept has been very inspiring.

References

Agostinho, A.A., L.C. Gomes and M. Zalewski. (2001) "The importance of floodplains for the dynamics of fish communities of the upper river Parana", in M. Zalewski, F. Schiemer, J. Thorpe, eds., *International Journal of Ecohydrology and Hydrobiology* 1: 1-2. Special issue on: Catchment processes land/water ecotones and fish communities. Warsaw 2001.

Gouder De Beauregard, A-C and G. Mahy. (2002) "Phytoremediation of heavy metals: the role of macrophytes in a stormwater basin", *International Journal of Ecohydrology and Hydrobiology* 2: 1-4. Proceedings of the Final Conference of the First Phase of the IHP-V Project 2.3/2.4 on Ecohydrology "The Application of Ecohydrology to Water Resources Development and Management" Venice, Italy, 16-18 Sep. 2001.

Hellegers, P.J.G.J and J-P.M. Witte. (2002) "Towards a simple integrated model for the re-wetting of nature reserves", *International Journal of Ecohydrology and Hydrobiology* 2:1-4. Proceedings of the Final Conference of the First Phase of the IHP-V Project 2.3/2.4 on Ecohydrology "The Application of Ecohydrology to Water Resources Development and Management" Venice, Italy, 16-18 Sep. 2001.

Holling, C.S., L.H. Gunderson, and C.J. Walters. (1994) "The structure and dynamics of the Everglades system: guidelines for ecosystem restoration", in S. Davis, J. Odgen, eds., *The Everglades: The Ecosystem and its Restoration*, Delray Beach, Florida: St. Lucie Press.

Hrbacek, J., M. Dvorakova, V. Korinek, and L. Prochazkova. (1961) "Demonstration of the effect of the fish stock on the species composition of zooplankton and the intensity of metabolism of the whole plankton association", *Verhandlungen der Internationale Vereinigung fur Theoretische und Angewandte Limnologie* 14: 192-195.

Jørgensen, S.E. (1996) "The application of ecosystem theory in limnology", *Verhandlungen der Internationale Vereinigung fur Theoretische und Angewandte Limnologie* 26: 181–192.

Mitsch, W.J. (1993) "Ecological Engineering - a co-operative role with planetary life-support system", *Environmental Science and Technology* 27: 438-445.

Ploskey, G.R. (1985) Impacts of terrestrial vegetation and preimpoundment clearing on reservoir ecology and fisheries in the United States and Canada. Food and Agriculture Organization of the United States, Fisheries Technical Paper 258, 1-35.

Reid, D.M and W.T. Momot. (1985) "Evaluation of pulse fishing for the walleye, Stizostedion vitreum vitreum, in Henderson Lake, Ontario", *Journal of Fish Biology* 27 (Suppl. A): 235-251.

Shapiro, J., V. Lamarra, and M. Lynch. (1975) "Biomanipulation: an ecosystem approach to lake restoration", in P.L. Brezonik, J. and L. Fox, eds., *Proceedings of a Symposium on Water Quality Management Through Biological Control*, University of Florida, pp. 85-96.

Timchenko, V. and O. Oksiyuk. (2002) "Ecosystem condition and water quality control at impounded sections of rivers by the regulated hydrological regime", *International Journal of Ecohydrology and Hydrobiology* 2: 1-4. Proceedings of the Final Conference of the First Phase of the IHP-V Project 2.3/2.4 on Ecohydrology "The Application of Ecohydrology to Water Resources Development and Management" Venice, Italy, 16-18 Sep. 2001.

Von Weizsaecker, E., A.B. Lovins, and L.H. Lovins. (1997*) Factor Four. Doubling Wealth-Halving Resource Use. The new Report to the Club of Rome,* London, U.K.: Earthscan Publishing Ltd.

Wagner-Lotkowska, I., Bocian, J., Pypaert, P., Santiago-Fandino, V., Zalewski, M., (2004) "Environment and economy - dual benefit of ecohydrology and phytotechnology in water resources management: Pilica River

Demonstration Project under the auspices of UNESCO and UNEP", *Ecology and Hydrobiology* 4 (3): 345-352.

Wagner I., Zalewski M., (2000) "Effect of hydrological patterns of tributaries on biotic processes in lowland reservoir: consequences for restoration", Special Issue. *Ecological Engineering* 16: 79-90

.Zalewski, M. and RJ. Naiman. (1985) "The regulation of riverine fish communities by a continuum of abioti-biotic factors", in J.S. Alabaster, ed., *Habitat Modification and Freshwater Fisheries*, London: FAO/UN/Butterworths Scientific, pp. 3-9.

Zalewski, M., B. Brewinska-Zaras, P. Frankiewicz and S. Kalinowski. (1990) "The potential for biomanipulation using fry communities in a lowland reservoir: concordance between water quality and optimal recruitment", *Hydrobiologia* 200-201: 549-556.

Zalewski, M., G.A. Janauer, and G. Jolankai. (1997) Ecohydrology. A new paradigm for the sustainable use of aquatic resources. UNESCO IHP Technical Document in Hydrology 7. IHP - V Projects 2.3/2.4, UNESCO Paris, pp. 58.

Zalewski, M. (2000) "Ecohydrology-the scientific background to use ecosystem properties as management tools toward sustainability of water resources. Guest Editorial", *Ecological Engineering* 16: 1-8.

Zalewski, M. (ed). (2002) Guidelines for the Integrated Management of the Watershed – Phytotechnology and Ecohydrology. United Nations Environment Programme, Division of Technology, Industry and Economics. International Environmental Technology Centre. Freshwater Management Series No. 5, pp. 188.

Zalewski, M. (2002) "Ecohydrology - the use of ecological and hydrological processes for sustainable management of water resources", *Hydrological Sciences Journal* 47: 825-834.

Zalewski, M. and R. Robarts. (2003) "Ecohydrology – a new Paradigm for Integrated Water Resources Management", *(Societas Internationalis Limnologiae) SIL News* 40:1-5.

Zalewski et al. (2004) "Integrated Watershed Management-Ecohydrology & Phytotechnology- Manual" (UNESCO, UNEP).

Section II:

Case Studies in Ecosystem-based Water Management

The Amazon Region

Luis E. Aragón

Center for Advanced Amazonian Studies/Federal University of Pará, Brazil

Introduction

The Amazon holds around 15% of the usable water on Earth, while other areas of the planet undergo severe shortages of water. What measures are needed in order to solve the problem and what is the role of the Amazon in this process? How can we reconcile the economic value and the vital character of water in a region that is extremely abundant of this resource? Up to recent years, technicians had quick answers for accessibility of water: build big dams and transfer water through aqueducts from humid areas to dry areas. Nowadays, such solutions are not easily accepted, as they are very costly from the environmental and economic points of view. As a result, solutions are being sought to diminish the demand, including the privatization of services to create a water market under the regulation of the World Trade Organization (WTO). Some commentators suggest the creation of an international water market through which countries with shortages of water could buy water from countries with abundance; or even the creation of an Organization of Water Exporting Countries. But can water be treated in similar terms to those adopted for other natural resources?

Fragile ecosystems such as the Amazon are regulated by the abundance of water in the hydrological cycle. Studies suggest that alteration of this cycle affects the entire life of the region, with serious climatic implications in other parts of South America and other continents (Salati, 1983). Public policies, however, are almost exclusively directed towards solving the problems of areas with scarcity of water, leaving aside problems related to conservation and management in areas of abundance. Thus, in dry areas, some privileged people have access to more water than they need, while in humid areas many underprivileged people suffer thirst, or sickness from consuming unsuitable water. In the final analysis, the need to manage access to water in humid areas, such as the Amazon, is as pressing as it is in dry areas. Contamination, lack of access to potable water, or bad management can generate ecological, economic and environmental disturbances with serious risks for the environment, health and human well-being. Although most environmental changes in the Amazon are natural processes, human intervention has accelerated those changes. This paper synthesizes conclusions of relevance to these problems in the management of water and ecosystems, based on research conducted by several scholars and published by UNESCO on Issues of local and global use of water from the Amazon under the auspices of the South-South Cooperation Programme for Ecodevelopment by UNESCO/MAB, UNU and the Third World Academy of Sciences (TWAS) (Aragón and Clüsener-Godt, 2004; 2003).

The Amazon Region

Amazonia is a complex region that presents numerous difficulties regarding its definition and analysis. The term Amazonia means so many different things, it is best understood as an immense area containing several Amazonias in the so called Greater Amazon Region, which includes the areas covered by the Amazon River basin and also the tropical rainforest (Gutiérrez Rey et al., 2004). Although there is no consensus regarding the exact extent of the region, it can be seen to cover approximately 8 million square kilometers, of which some 6,878,000 belong to the Amazon River basin (Dominguez, 2004). Approximately 28 million people were estimated to be living in the Greater Amazon Region in the year 2000, including about 1 million Amerindians. More than 60% of the current population lives in urban areas. Eight countries and a French Department share the region: Bolivia, Brazil,

Colombia, Ecuador, Guyana, Peru, Suriname, Venezuela and French Guyana (Aragón, 2005a).

Figure 1: The Amazon.

The Amazon Region is one of the largest, most diverse, complex and rich natural domains of the planet. The area of the entire Amazon Region corresponds to 1/20 of the surface of the earth, 2/5 of South America, 3/5 of Brazil; it contains 1/3 of the tropical forest, and only a tiny portion of the world population (Becker, 1990). The relief includes valleys, plateaus (Brazil and Guyanas) and high mountains (Andes); and the hydrography includes rivers, lakes and the Atlantic coast. About 30% of all animal and plant species of the world are found in the Amazon region. The diversity existing in the Amazon is expressed in different ways. In the Peruvian Amazon, for example, the largest concentration of species of trees in the world has been registered: 300 species per hectare; in 2km^2 630 species of vertebrates were found, including 353 species of birds, and 232 species of butterflies; and 5,000 species of insects in just one tree (CDEA, 1992).

Huge abundance of fresh water, thick tropical rainforest, and biological and cultural diversity are common features of this enormous region; and its functioning is intimately related and dependent of each one of those factors. The river system, Amazonas-Solimões-Ucayalli, represents the longest river in the world, with 6,671 kilometers and the whole basin is constituted by more than 1,000 rivers (CDEA, 1992). As the Amazon River approaches the ocean, the sediments accumulate to an estimated quantity of 1 billion tons per year, which the river discharges into the ocean (Botto, 1999). Such sediments are dispersed along the coast up to the Orinoco Delta, making this area rich in soils for agriculture (Lima, Tourinho and Costa, 2000). But this is also a very fragile ecosystem, with a diverse fauna and flora that is dependent on the mixture of sea and fresh water (Prost and Mendes, 2001).

Available Water Resources

Why is there so much water in the Amazon and what is its origin? The existence of water in the Amazon is due to the geological history of the Earth and its global hydrological cycle. However, the amount and distribution in the region is also related to regional and local factors. The Earth's hydrological cycle incorporates several elements, including energy exchange, transportation of water vapor, precipitation, drainage, infiltration, and a variety of water storage mechanisms; all of which depend on weather, ecosystems and other factors. Water vapor in the atmosphere regulates Earth's temperature by a natural greenhouse effect so that temperatures near the surface are around 30°C beyond what would be expected purely from radioactive effects of solar energy on the Earth, without the humidity of the troposphere (Souza, Rocha and Cohen, 2004).

The Amazon Region is located within the Inter Tropical Convergence Zone (ITCZ), a high precipitation latitudinal zone around the equator where vapor produced in neighboring zones converges at low levels and rises, producing widespread cloud coverage, and a predominance of precipitation over evaporation. This leads to large volumes of precipitation throughout the year. The Andean mountain range interferes in the longitudinal distribution of water by blocking the transportation of humid air from the Atlantic Ocean consequently further increasing precipitation rates on the Amazonian side of the mountains (Souza, Rocha and Cohen, 2004).

From the bio-geophysical point of view, the Amazon water cycle has been considered to have been in balance over the last few centuries. How does such a balance work? The information needed to understand how local hydrological resources are maintained or changed includes data on spatial distribution of vegetation types, liquid surfaces, soil moisture contents, surface runoff, subterranean water flows, air humidity and temperature, precipitation, and other gas fluxes to the atmosphere. Research is ongoing on these issues and although much research is still needed to fully understand the determining factors of the water cycle in the Amazon, at the regional level, some conclusions have already been reached. These include the following:

- Fifty percent of the water vapor existing in the Amazon is transported westward by the winds coming from the Atlantic Ocean; the other 50% comes from evapotranspiration of the forest itself (Salati, 1983).

- Precipitation varies in the region for reasons including the position of the barrier of the Andean Mountains, and the location of the Region in the Northern as well as in the Southern Hemisphere, which produces a difference between the rainy and dry seasons (Souza, Rocha and Cohen, 2004).

- The waters in the region vary with regard to the levels of sediments and nutrients that they contain. The so-called black waters are extremely poor in nutrients and sediments, in relation to white waters. This is the result of the geological characteristics of the places of origin of the rivers, the types of soils in the river basins and the types of vegetation (Junk, 1983).

Current Basic Human Needs for Water and Relationships to Human Well-being

From the above synthesis, it is possible to conclude that the Amazon does not present a deficit of water for human needs. Out of 180 countries and territories in the world, the Amazonian countries rank among the 33 most water abundant (UNESCO-WWAP, 2003). However, when it comes to safe access to this water for drinking, according to the UNDP index of human development (UNDP, 2004), the Amazonian countries are ranked between the 67th (Suriname) and the 114th (Bolivia) position; the access to improved water sources goes from 80% of the population in Peru to 94% in Guyana, and to sanitation services from 93% in Suriname to 70% in Bolivia. Furthermore, while those figures reflect principally the condition in urban areas, they hide significant regional differences. In Brazil, for example, in 2000, only 62.5% of households located in urban areas of the Amazon (North region), were served with water, while in rural areas, those services are even lower than they are in urban areas, and quite often non-existent (UNICEF, 2004). In comparison, several countries from other regions with severe limitations regarding the available quantity of water (less than 1,000 cubic meters per person per year) present similar or higher percentages of access to water and sanitation services than the well-endowed Amazonian countries (Table 1).

Table 1: Water availability and access to water and sanitation in Amazonian countries and selected countries with scarcity of water.

Country	Human Development Index* (2002)		Water Availability**		%Population with sustainable access to* (2000)	
	Rank	Index	Rank	M^3/p/ year	An improved water source	Improved Sanitation
Amazonian countries						
Suriname	67	0.780	6	229,566	82	93
Venezuela	68	0.778	23	51,021	83	68
Brazil	72	0.775	25	48,314	87	76
Colombia	73	0.773	24	50,635	91	86
Peru	85	0.752	17	74,546	80	71
Ecuador	100	0.735	33	34,161	85	86
Guyana	104	0.719	5	316,689	94	87
Bolivia	114	0.681	16	74,743	83	70
French Guyana	-	-	3	812,121	-	-
Selected countries with less than 1,000 m^3 of water per person per year						
Libya	58	0.794	174	113	72	97
Saudi Arabia	77	0.768	173	118	100	95
Jordan	90	0.750	170	179	96	99
Tunisia	92	0.745	162	482	80	84
Algeria	108	0.704	163	478	89	92
Egypt	120	0.653	156	859	97	98
Morocco	125	0.620	155	971	80	68

Sources: * UNDP – United Nations Development Programme. (2004) Human Development Report 2004. New York: UNDP.
** UNESCO-WWAP. (2003) Water for people, water for life: World water development report. Barcelona: UNESCO and Berghahn Books.

Pressures on Water Regulation by Ecosystems

Ecosystems in the Amazon depend on abundance of water and the functioning of the hydrological cycle. Although the hydrological cycle of the Amazon is considered to be in equilibrium, serious disturbances have been introduced by human activities. These disturbances indicate alterations to the cycle occurring initially on a local scale. The following are examples of such disturbances:

According to Souza, Rocha and Cohen (2004), deforestation and substitution of vegetation affect the water cycle (among other things) by:

- reduction of the retention of humidity in the soil's top layer;
- facilitation of sudden evaporation of water previously retained in the forest canopy, producing as a consequence a new balance in surface radiation;
- increased reflectivity (albedo) and temperature;
- convection and the formation of shallow cumulus clouds that usually do not evolve into nimbus clouds, and thus may not produce rain;
- medium- and long-term decreases in humidity in both soil and air, resulting in a decrease in the concentration of condensation nuclei that are needed to form cumulonimbus clouds.

The common slashing and burning processes used in the region suddenly increase the concentration of aerosols from levels estimated to be around 200cm^{-3} to 20,000 cm^{-3}. This increase in the concentration of aerosols per cubic centimeter in the air during biomass burning, together with the heat generated by the fires, generates an immense evaporation and convection of air mass from the surface, causing, in the short run, more potential cloud coverage and precipitation. By the end of August 2002, the NOAA-12 Satellite had detected around 10,300 fire sites occurring every month in the Amazon region. The state of Mato Grosso in the Brazilian Amazon registered 26,400 fires in the first eight months of that year. It is estimated that 900,000 tons of aerosols and gases are sent into the atmosphere annually in the Mato Grosso region alone. With such burning activity taking place, scientists believe that the release of monoxide, dioxide and other gases into the atmosphere could increase the greenhouse effect in the region (Souza, Rocha and Cohen, 2004).

Serious damage to the forest and soils is anticipated. Bare soil loses porosity through compaction, and increased rainfall may cause faster rainfall drainage, erosion and silting of the rivers and banks (Souza, Rocha and Cohen, 2004). Studies revised by Souza, Rocha and Cohen (2004) have demonstrated that even when soil is regenerated by small bushes or grass, significant micro-climatic changes in the surface emerge. Comparing forest and pasture areas in the Brazilian Amazon, scholars found the following results:

- forest areas absorbed 11% more solar radiation than pasture areas;
- average albedo in forest areas was 13.4% and in pastures areas was 18%;
- average temperature during the day on the soil of forest areas was around 24.1°C, while in the pastures areas was 33°C;
- daily temperature of soil varies with depth; at 20cm depth, in forest areas such variation did not exceed 2.8°C, but in the pasture areas the variation was of 8°C;
- volumetric moisture contents within the uppermost meter of soil beneath pastures areas were generally 15% smaller than under nearby forests at the same depth interval. At about 2 meters, the soil moisture beneath pastures may exceed

the corresponding values of $0.36m^3/m^3$ found under the forests. Down to 4 meters depth, the forest roots and leaf systems are always more effective in pumping water to supply the transpiration of vegetation, than pastures.

These results led to the conclusion that the removal of the forest would reduce the level of humidity in the air above ground (estimated 20 to 30%), and that large-scale deforestation would reduce precipitation by 5 to 20%.

Trends in Habitat Destruction & Water Availability

Regardless of the efforts of governments, NGOs and other actors, deforestation in the Amazon continues, especially in the Brazilian Amazon. In June 2003, the Brazilian National Institute of Spatial Research (INPE) announced, that $25{,}500km^2$ of the Brazilian Amazon had undergone deforestation between 2001 and 2002, representing an increase of 40% in the rate of deforestation over the previous period. The reaction to this announcement from scientists, NGOs, the media, and even politicians was of indignation; debates, conferences, discussions and other events occurred at different levels and parts of the country. Shortly afterward, the government announced a Plan of Action to Prevent and Control Deforestation in the Amazon (PR, 2004). According to INPE's projections, the acceleration of deforestation elevated the total deforested area of the Brazilian Amazon to 631.369 km^2, or 15.7% of the total area of the region in 2002 (PR, 2004). Some 25% of this area is abandoned or sub-utilized and in many cases degraded. Nevertheless, despite the expectations of the government, deforestation of the Brazilian Amazon continued to increase. According to INPE 27, $362km^2$ were deforested between 2003 and 2004, 43% of which within the State of Mato Grosso, one of the largest soybeans producer states in the country (INPE, 2006). This figure elevated the total deforested area of the Brazilian Amazon to about $680{,}000km^2$, or approximately 17% of the region. Numeric models suggest that large-scale deforestation in the Amazon can diminish superficial drainage between 10 and 20% and increase the temperature of the air near the surface between 0.6 and 2.0°C. This prolongs the periods of drought in the region and, in consequence, alters the regional water cycle (Souza, Rocha and Cohen, 2004; Dias, 2003; Souza, 2003).

Deforestation is associated with increased fires, and the expansion of soybean cultivation, cattle raising, and illegal timber exploitation (Aragón, 2005b). Other environmental risks are contamination of water, especially in large cities of the region where there is a lack of sanitation facilities, and also around mining areas. Agro toxics, mercury, arsenic, and other contaminating substances have been found in the Amazonian rivers (Fenzl and Mathis, 2004; Ruivo, 2003; Braz, 2003). With the increase of population, urbanization, building of roads, dams and infrastructure complexes, extensive agriculture, oil exploitation, mining (especially by using mercury) and implementation of unsustainable development plans, disturbances in the Amazonian ecosystems have been intensified at the local level. These disturbances affect biodiversity, micro climates and quality of water resources, and potentially intervene in the hydrological cycle of the region.

Drivers of Change: Population, Development Patterns, Increased Demand and Environmental Change

Although the population has increased in the Amazon region, and is growing at higher rates (around 3 percent per year) than in the various countries, population numbers do not represent a problem (Aragón, 2005a). Considering the amount of water existing in the region, even if the demand for water increase in the future, the supply will remain sufficient. The issue here is to allow access to potable water and sanitation services to all inhabitants of the region. Environmental education and management are also extremely important in the Amazon, considering the misperception of the population who perceive water as an unlimited resource, when observing the hugeness of rain, humidity, rivers and lakes. Studies have already demonstrated the death of rivers in cities, the contamination of fish and humans by mercury and the rise in contamination of river beaches, rendering them inappropriate for human use during certain periods of the year (Fenzl and Mathis, 2004; Braz, 2003).

Even though environmental awareness has increased in the world, reaching the highest levels of governments, plans for sustainable development in the Amazon region have not yet fulfilled the expectations placed upon them (Aragón, 2002). Allied to the increase of deforestation, are the facts of accelerated urban growth, expansion of agriculture and industrialization, all of which place increased pressures on consumption of water. Even recognizing that many cities in the world, especially those located in developing countries, are already suffering water crises, those places are the ones with better access to potable water in relation to poor rural areas. This is also true in the Amazonian countries. Therefore, the demand for water tends to increase with more people living in cities. In addition, the so-called productive use of water (in agriculture and industry) is increasing, as industrialization progresses and the demand for agricultural products increases. So the competition among different uses of water has become more intense, affecting the accessibility for human consumption, and at the same time raising the economic value of water.

Integrated Management of Water Resources, Ecosystems, Human Well-Being and Ecosystem Services – Current Progress and Problems

It is clear that issues related to the use of water in the Amazon are mainly related to management and public policies since:

- there is no lack of fresh water resources in the region;
- the region does not need to waste water with irrigation;
- more than 80% of the forest is still preserved or has suffered little intervention;
- most of the rivers of the basin cross two or more countries, but there are no serious conflicts between the countries;
- there is no problem with population numbers. The population density of the region is around 4.0 people per square kilometer (Aragón, 2005a; Becker, 2004).

In spite of these natural advantages, and although all Amazonian countries have legislation concerning the use and management of water, public policies are mainly directed towards issues related to arid areas. In Brazil, the law considers the basin as the management unit, and advocates for decentralization and the participation of local communities (Setti, 2004). These principles have been applied in some areas, but very little in the Amazon.

It is clear then, as Becker (2004) pointed out, that issues of local use of water in the Amazon are very specific. While global problems are mainly characterized by lack of supply or availability and great increase in consumption, in the Amazon the problems are related mainly to solving the paradox of abundance of water and limited accessibility principally because of the lack of access to water distribution services and sanitation facilities.

What measures are needed to overcome those problems in order to make the best possible use of water of the region for the well-being of the people of the region, the Amazonian countries and the world? New strategies of management are needed in these areas to tackle specific local problems. Although all Amazonian countries have their own national systems of water resource management, there are few explicit measures for humid areas. In Brazil, for example, the National Water Agency (ANA) was created in 2000 to reinforce the law on water resources that was approved in 1997 (Setti, 2004). This law determines, among other things:

- water is a public good;
- water is a limited natural resource which has economic value;
- when there is shortage of water resources, priority is given to human consumption and the provision of water for animals;
- the management of water resources must always allow for the multiple uses of water;
- the water basin is the territorial unit for the implementation of policies and management;
- water resource management must be decentralized and should involve the participation of the government and that of users and communities.

Nevertheless, in spite of those regulations, the Amazonian countries tend to follow the world tendency of privatizing water supply services. Many justify this trend because of the limited supply of this natural resource. It is thought that individuals are more inclined to save water if they are made to pay for it. Although this idea is open to debate as concerns irrigation, and the argument may be valid for the use of water for industrial purposes, it is contentious to argue about the commercialization of potable water for human consumption and domestic uses. For ethical reasons, access to water must be guaranteed to all human beings. It is unacceptable in the 21st century that thousands, even millions of people do not have access to potable drinking water, even in areas extremely abundant of this resource, and that so many children die due to consumption of unsuitable water (Dias and Aragón, 2004).

There exists strong debate concerning that issue. In Cochabamba (Bolivia), just to mention a case in an Amazonian country, Bechtel, a large San Francisco based firm, received a concession to operate water services. In December 1999, the firm doubled the price charged for water. The population protested, which led to the death of some demonstrators. The government revoked its water privatization legislation and Bechtel is now suing the Bolivian government for 40 million dollars (Dias and Aragón, 2004). Today the whole country is in a political turmoil demanding nationalization of enterprises and services.

Production and exportation of energy is another big issue in the Amazon. The questions surrounding hydroelectric dams are: -why? -for whom? and, -at what cost? The immense hydrological potential of the Amazon is well recognized. Several dams have already been constructed and others are planned in the Brazilian Amazon. This energy serves many places and industries located outside the region, while the region itself remains deficient in electrification, especially in rural areas. Some advocate compensation payments for the construction of such dams, others argue to look for other sources of energy, considering the high environmental costs involved in building dams in the region (Machado and Souza, 2004; Rocha, 2003).

Capacity Development: Institutional, Managerial, Human, Technological Capacity

Building regional capacity at all levels will be one of the key elements to overcome problems related to accessibility of potable water, sanitation and the improvement of quality of life in the Amazon region. Although environmental awareness is increasing in the Amazonian countries, the challenges of implementing policies for sustainable development have been great. Even if the Amazonian countries manage to achieve the Millennium Development Goals on time, the Amazon region is unlikely to do so (UN Millennium Project, 2005), in spite of recent efforts made in that direction. The following paragraphs present a brief overview of these efforts.

In 1978, the Amazonian countries signed the Amazon Cooperation Treaty "to undertake joint actions and efforts to promote the harmonious development of their respective Amazonian territories in such a way that these joint actions produce equitable and mutually beneficial results and achieve also the preservation of the environment and the conservation and rational utilization of the natural resources of those territories" (MRE, 1978). The Amazon Cooperation Treaty Organization (OTCA) was created in 2000 and reinforced by the parliaments of all countries in 2002, as a multilateral agreement with a permanent Secretariat in Brasília, responsible for the formulation and coordination of the Program of the Treaty. This Organization represents the ideal instrument for identifying the true potential of the region and to formulate and implement programs and actions that serve the entire Amazon and should lead to sustainable development. OTCA will also serve as a catalytic organization for actions carried out by NGOs, universities, research institutes and governmental organizations at all levels. It is the best attempt at sub-regional integration in the history of the Amazon, enabling countries to agree on the principles that will guide the development of the region. Many rivers of the Amazon basin are international rivers, so legislation, use, management, pollution control, transportation, and many other issues related to water in the region should be addressed by OTCA. In 2004, all countries approved the Strategic Plan of the Treaty for 2004-2012 (OTCA, 2004). This program reflects a new approach to valuing the forest, biodiversity, cultural diversity, water resources and many other factors in search of sustainable development.

One of the main areas of the program refers to water resources. In this regard, OTCA has been implementing an ambitious GEF Project on Integrated and Sustainable Management of Transboundary Water Resources in the Amazon River Basin since June 2005. "to strengthen the institutional framework for planning and executing, in a coordinated and coherent manner, activities for the protection and sustainable management of the land and water resources of the Amazon River Basin in the face of ongoing climatic changes being experienced in the Basin" (OTCA, 2006).

Parallel to OTCA's activities but coinciding in objectives, other initiatives are underway in the Amazon to build regional capacity (Aragón, 2005c). Among these initiatives are those implemented through the Association of Amazonian Universities (UNAMAZ) involving all the Amazonian countries in the areas of science and higher education (Acevedo, 2003). It is important also to emphasize that the number of graduate programs addressing the environmental aspects of the Amazon has increased over the last few years. Amongst these, the graduate program on sustainable development of the humid tropics of the Center for Advanced Amazonian Studies of the Federal University in Belém focuses on the whole Amazon region. The masters and doctorate degree programs on Fresh Water Biology and Interior Fishing offered by the National Institute for Amazonian Research in cooperation with the Federal University of Amazonas in Manaus, also offer great potential for expansion to all Amazonian countries. Another example is the Masters course on Coastal Ecosystems at the Federal University of Pará in the city of Bragança. Also, the Center for Development Studies of the Central University of Venezuela in Caracas established a Masters course on health and environment based upon Amazonian issues. In terms of research, the Program for Tropical Coastal Ecosystem Studies (ECOLAB), which gathers researchers from institutions in Suriname, French Guyana, and Amapá, Pará and Maranhão states in Brazil to study and monitor environmental change along the Amazonian coast (Prost, 2003). Other relevant programmes include: Processes of Change in the Amazon Estuary due to Anthropogenic Activities and Environmental Management (MEGAM), conducted by the Center for Advanced Amazonian Studies of the Federal University of Para; the Program on Management and Dynamics in Mangrove Swamp Areas in Northeastern Para (MADAM), developed by the Federal University of Para at Bragança; and the Program on Natural Resources and Anthropology of Maritime, Riverbank and Estuarine Societies (RENAS), coordinated by the Goeldi Museum in Belém.

Ongoing capacity building activities are strengthened by networks and exchanges such as the Amazonian Initiative, a network of institutions of the Amazonian countries gathered to study and develop agricultural and related sciences (Aragón, 2005b); the South-South Cooperation Program for Ecodevelopment implemented jointly by UNESCO/MAB, UNU and TWAS which has supported important projects in the region such as the use of water from the Amazon, management of coastal ecosystems, comparative research on agriculture in the humid tropics, comparative analysis of Biosphere Reserves in the Amazon and other Humid Tropical Areas, and the study of population dynamics and its relation to environmental change in the Amazon (Clüsener-Godt, 2004); the Large-Scale Biosphere Experiment in Amazonia (LBA), an international research initiative coordinated by the National Institute of Amazonian Research of the Ministry of Science and Technology of Brazil (INPA) in Manaus. This initiative is designed to create the knowledge needed to understand the climatic, ecologic, biogeochemical, and hydrological functioning of Amazonia, the impact of land use change on these functions, and the interactions between the region and the Earth system. LBA is sponsored by NASA and other agencies and includes more than 240 research and higher education institutions from Brazil, other Amazonian countries, Europe and the United States totalizing more than 1600 scholars and students and more than 120 research projects. Of the total institutions involved in the LBA Program, more than 100 are Brazilians, and more than 40 are located in the Amazon region (Luizão, 2005).

Conclusions

Because this region possesses immense quantities of fresh water, biodiversity, forest, sun energy, and winds, the sovereignty of the Amazonian countries over the region has been called into question in the name of the common good, or the environmental health of the Planet (Becker, 2004; Costa, 2003). In the final analysis, the critical issue concerns rights to water. On the one hand, this resource is frequently seen as an economic good, to be regulated by the market (Castro, 2003). On the other hand, the debate continues as to whether such a rare resource in the world as fresh water should be treated as a public good, belonging to humanity, or whether it should be subject to commercial rules. Where is the limit between ethics and economics?

Concerning the Amazon, it is essential to recognize that issues related to the use of water in the region are very specific. Therefore, special measures to attend the local needs in humid regions are necessary. How to deal with problems related to abundance of water in tropical areas such as (among others) the spread of tropical diseases that occur and disseminate at a very high speed (Yarzábal, Espinal and Aragón, 1992); floods that kill many and produce large economic damages especially in highly populated tropical humid countries; environmental changes affecting precipitation and climate in the Amazon and other areas; and improvement of water quality for consumption (Fenzl and Mathis, 2004). The main problem in the region, however, concerns accessibility and not availability of the resource. At the end of the day, the question for debate is how to use water from the Amazon for human well-being without depriving local population needs, the environment and the sovereignty of the countries sharing this region.

References

Acevedo, R.E.M. (2003) UNAMAZ: Um projeto de cooperação pan-amazônica. Belém: UNAMAZ.

Aragón, L.E. (2002) "Há futuro para o desenvolvimento sustentável na Amazônia?", in A. Fiúza de Mello, ed., *O futuro da Amazônia: Dilemas, oportunidades e desafios nolimiar do Século XXI*, Belém: UFPA, pp. 33-53.

Aragón, L.E. (2005a) "Até onde vai a Amazônia e qual é a sua população?", in L.E. Aragón, ed., *Populações da Pan-Amazônia*, Belém: NAEA, pp. 13-23.

Aragón, L.E. (2005b) "Building regional capacity for sustainable use of tropical rainforest in Latin América: Experience and challenges of the South-South cooperation", in M. De Dapper, ed., *Tropical Forests in a Changing Global Context*, Brussels: Royal Academy of Overseas Sciences/UNESCO, pp. 257-265.

Aragón, L.E. (2005c) "Cooperação Sul-Sul para o desenvolvimento científico e tecnológico da Amazônia", *Parcerias Estratégicas* 20: 767-794.

Aragón, L.E. and M. Clüsener-Godt (eds). (2004) *Issues of local and global use of water from the Amazon*, Montevideo: UNESCO, NAEA/UFPA.

Aragón, L.E. and M. Clüsener-Godt (eds). (2003) *Problemática do uso local e global da água da Amazônia*, Belém: UNESCO/NAEA.

Becker, B. (1990) *Amazônia*, São Paulo: Ática.

Becker, B. (2004) "Inclusion of the Amazon in the geopolitics of water", in L.E. Aragón and M. Clüsener-Godt, eds., *Issues of local and global use of water from the Amazon*, Montevideo: UNESCO, NAEA/UFPA, pp. 143-166.

Botto, M.P. (1999) "The Amazon Cooperation Treaty: A mechanism for cooperation and sustainable development", in Asit K Biswas et al., eds., *Management of Latin American River Basins: Amazon, Plata and São Francisco*. Tokyo, Japan: United Nations University Press, pp. 68-100.

Braz, V.N. (2003) "Poluição dos recursos hídricos da Amazônia: Aspectos urbanos", in L.E. Aragón, and M. Clüsener-Godt, eds., *Problemática do uso local e global da água da Amazônia*, Belém: UNESCO/NAEA, pp. 153-158.

CDEA (Commission on Development and Environment for Amazonia). (1992) *Amazonia without myths*, Washington, D. C.: IDB/UNDP/ACT.

Castro, E. (2003) "Geopolítica da água e novos dilemas à propósito da Amazônia e seus recursos naturais", in L.E. Aragón and M. Clüsener-Godt, eds., *Problemática do uso local e global da água da Amazônia*, Belém: UNESCO/NAEA, pp. 321-342.

Clüsener-Godt, M. (2004) "Implementing Agenda 21: Nature conservation and sustainable development in the humid tropics", *Tropical Ecology* 45(1): 183-186.

Costa, W.M. (2003) "Valorizar a água da Amazônia: Uma estratégia de inserção nacional e internacional", in L.E. Aragón and M. Clüsener-Godt, eds., *Problemática do uso local e global da água da Amazônia*, Belém: UNESCO/NAEA, pp. 299-320.

Dias, N.L. (2003) "Comentários sobre "impactos antropogênicos no ciclo hidrológico da Amazônia," de Souza, Rocha e Cohen", in L.E. Aragón and M. Clüsener-Godt, eds., *Problemática do uso local e global da água da Amazônia*, Belém: UNESCO/NAEA, pp. 95-105.

Dias, M.A. and L.E. Aragón. (2004) "Amazonian co-operation for knowledge on water resources and for the sustainable use of these resources in the region", in L.E. Aragón and M. Clüsener-Godt, eds., *Issues of local and global use of water from the Amazon*, Montevideo: UNESCO, NAEA/UFPA, pp. 219-237.

Dominguez, C. (2004) "The importance of rivers for the transportation system of the Amazon", in Luis E Aragón and Miguel Clüsener-Godt, eds., *Issues of local and global use of water from the Amazon*, Montevideo: UNESCO, NAEA/UFPA, pp. 77-100.

Fenzl, N. and A. Mathis. (2004) "Pollution of natural water resources in Amazonia:Sources, risk and consequences", in L.E. Aragón and M. Clüsener-Godt, eds., *Issues of local and global use of water from the Amazon*, Montevideo: UNESCO, NAEA/UFPA, pp. 57-76.

Gutiérrez Rey, F., L.E. Acosta Muñoz and C.A. Salazar Cardona, (2004) *Perfiles urbanos en la Amazonia Colombiana: Un enfoque para el desarrollo Sostenible*, Bogotá: Instituto Sinchi.

INPE (Instituto Nacional de Pesquisas Espaciais). (2006) *Monitoramento da floresta amazônica brasileira por satélite. Projeto PRODES* available from www.obt.inpe.br/prodes/index.html.

Junk, W.J. (1983) "As águas da região amazônica", in E. Salati, H.O.R. Shubart, W. Junk and A.E. de Oliveira, eds., *Amazônia: Desenvolvimento, integração e ecologia*, Belém: Museu Paraense Emílio Goeldi, pp. 45-100.

Lima, R.R., M.M. Tourinho and J.P.C. da Costa, (2000) *Várzeas Flúvio-Marinhas da Amazônia: Características e possibilidades agropecuárias*, Belém: Faculdade de Ciências Agrárias do Pará.

Luizão, F.J. (2005) "Interações biosfera-atmosfera na Amazônia: contribuições do projeto LBA ao conhecimento e ao desenvolvimento sustentável da região", *Parcerias Estratégicas* 20: 681-708.

Machado, J.A. da Costa and R.C.R. Souza. (2004) "Determining factors in the construction of hydroelectric power plants in the Amazon: Reasons for claiming damage payments", in L.E. Aragón and M. Clüsener-Godt, eds., *Issues of local and global use of water from the Amazon*, Montevideo: UNESCO, NAEA/UFPA, pp. 101-142.

MRE (Ministério das Relações Exteriores/Brasil). (1978) Tratado de Cooperação Amazônica. Brasília, D.F.: MRE.

OTCA (Amazon Cooperation Treaty Organization). (2004) Strategic Plan of the Amazon Cooperation Treaty Organization (2004-2012). DOC/XII ACC-ACTO/04. Brasília, D. F.: OTCA.

OTCA (Amazon Cooperation Treaty Organization). (2006) *Global Environment Facility – Project Document. Integrated and Sustainable Management of Transboundary Water Resources in the Amazon River Basin* available from http://www.otca.br/arquivosdoc/projetogef.pdf.

PR (Presidência da República/Casa Civil). (2004) Plano de ação para a prevenção e controle do desmatamento na Amazônia Legal. Brasília, D. F.: Grupo Interministerial para a redução dos índices de desmatamento da Amazônia Legal.

Prost, M.T. (2003) "ECOLAB – Laboratório de ecossistemas costeiros", in L.E. Aragón, ed., *Conservação e desenvolvimento no estuário e litoral amazônicos*, Belém: NAEA/UNESCO, pp. 69-82.

Prost, M.T. and A.C. Mendes (eds). (2001) *Ecossitemas costeiros: Impactos e gerenciamento ambiental*, Belém: Museu Paraense Emilio Goeldi.

Rocha, G. de Miranda. (2003) "Usinas hidrelétricas, apropiação dos recursos hídricos na Amazônia e desenvolvimento regional", in L.E. Aragón and M.

Clüsener-Godt, eds., *Problemática do uso local e global da água da Amazônia*, Belém: UNESCO/NAEA, pp. 253-269.

Ruivo, M. de Lourdes Pinheiro. (2003) "Uso da água na Amazônia: Fontes de poluição, riscos, conseqüências e alternatives", in L.E. Aragón and M. Clüsener-Godt, eds., *Problemática do uso local e global da água da Amazônia*, Belém: UNESCO/NAEA, pp. 143-151.

Salati, E. (1983) "O clima atual depende da floresta", in E. Salati, H.O.R Shubart, W. Junk and A.E de Oliveira. *Amazônia: Desenvolvimento, integração e ecologia*, Belém: Museu Paraense Emílio Goeldi, pp. 15-44.

Setti, A.A. (2004) "Legislation and institutional systems for water resources management in Brazil and their relevance to the Amazon", in L.E. Aragón, M. Clüsener-Godt, eds., *Issues of local and global use of water from the Amazon*, Montevideo: UNESCO, NAEA/UFPA, pp. 167-201.

Souza, E.L. de. (2003) Comentários sobre "Avaliação dos impactos antropogênicos no ciclo da água na Amazônia," de Souza, Rocha e Cohen in L.E. Aragón, M. Clüsener-Godt, eds., *Problemática do uso local e global da água da Amazônia*, Belém: UNESCO/NAEA, pp. 107-114.

Souza, J.R.S., E.J.P. Rocha and J.C.P. Cohen. (2004) "Evaluation of the anthropogenic impacts on the water cycle in Amazonia", in L.E. Aragón, M. Clüsener-Godt, eds., *Issues of local and global use of water from the Amazon*, Montevideo: UNESCO, NAEA/UFPA, pp. 37-56.

UNESCO-WWAP. (2003) Water for people, water for life: World water development report, Barcelona: UNESCO and Berghahn Books.

UN Millennium Project. (2005) *Health, Dignity, and Development: What will it take?* Task Force on Water and Sanitation.

UNDP (United Nations Development Programme). (2004) Human Development Report-2004, New York: UNDP.

UNICEF (United Nations Children's Fund). (2004) Ser criança na Aamazônia: Uma análise das condições de desenvolvimento infantil na Região Norte do Brasil, Belém: UNICEF.

Yarzábal, L., C. Espinal, and L.E. Aragón (eds). (1992) *Enfoque integral de la salud humana en la Amazonia*, Caracas: UNAMAZ.

The Gulf Region

Ian McCann

University of Delaware, USA

Introduction

The Gulf is a term commonly used to refer to the countries of the Arabian Peninsula that share, with Iran, the body of water known both as the "Persian Gulf" and the "Arabian Gulf", or more simply just "the Gulf" (Figure 1). It is used in this context to include the Arabian Peninsula countries that are members of the Gulf Cooperation Council (GCC), namely Kuwait, Saudi Arabia, Bahrain, Qatar, United Arab Emirates, and Oman. They all have the benefit of oil resources to varying degrees, and the wealth generated from exploiting this resource has produced one of the most rapid economic development rates in history. Geographically, Saudi Arabia dominates by virtue of being larger than all the others combined, and is bounded not just by the Gulf on the East, but by the Red Sea on the West. Oman is the second largest of the countries, but technically is bounded by the Gulf of Oman rather than the Gulf (except across the strategic Strait of Hormuz where the Musandam enclave is located). Yemen is not currently a member of the GCC, although its membership is under negotiation. Iraq technically is also connected to the Gulf by the Shatt al Arab waterway, but is not a member of the GCC and has very different water resources by virtue of the Tigris and Euphrates rivers that flow through it.

Figure 1: Map of the Gulf.

Condition and Trends of Water Resources in the Gulf

The Gulf is one of the most arid regions in the world, with very low precipitation and very high evaporation rates. There are substantial reserves of fossil groundwater, particularly in Saudi Arabia, and these are being exploited for irrigated agricultural production. These resources are being rapidly depleted, with the water remaining projected to last as little as 20 years in some locations if current withdrawal rates continue. Renewable water resources, particularly of good quality water, are very limited (Zubari, 1998; Al-Rashed and Sherif, 2000; Böer, 1997). Desalination is increasingly becoming the primary source of domestic and municipal water. There is also increased use of treated waste water (recycled water).

The rapid economic development made possible by income from oil production has profoundly changed the Gulf. The human population has grown significantly, (Figure 2) and is no longer limited by natural resource, ecological or other constraints. Income per capita and the standard of living of the citizens has increased dramatically, and some of the countries are now among the worlds most advanced with respect to these indicators. Food availability, in quantity, quality and variety, both internally produced and imported, has likewise increased tremendously. The availability of infrastructure such as communications and transportation, and access to health care, education and other services has also greatly improved.

Figure 2: Human Population of Gulf Countries.

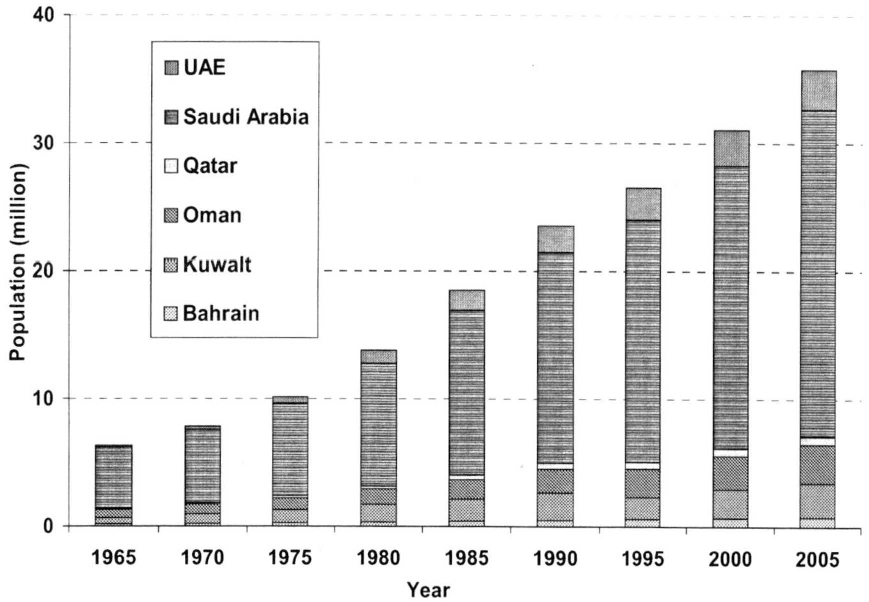

The energy resources of the Gulf have enabled the countries to meet their domestic and municipal water requirements using seawater desalination. The region is the largest producer of desalinated water in the world, and output continues to grow to meet increasing demand. Desalination plants are often part of oil or gas powered electrical power plants, and they use the waste heat from power generation in the desalination boilers. The plants are located along the coast, and discharge concentrated brine back into the Gulf as a waste stream. Desalinated water is generally supplied to the citizens of the Gulf countries at subsidized cost, and so it is relatively inexpensive. This has tended to cause consumption to increase, and in some countries per capita water consumption is now higher than in many other countries with much larger water resources. Indeed, perhaps desalination has come to be regarded as a quasi-renewable water resource because the substantial energy reserves of the region will allow it to continue into the foreseeable future. The cost of desalination has also decreased in recent years, although it is unlikely to ever be economical as a water source for extensive irrigated field agriculture because of the large quantities required. For example, one hectare of fully irrigated alfalfa requires 30,000 to 40,000 m^3 of water per year, and the cost of desalinated water is unlikely to decrease sufficiently to make this economically feasible. Desalinated water can however probably be justified for the production of high value crops such as greenhouse vegetables, fruits and flowers.

Water Resource Use and Regulation by Ecosystems

This paper illustrates two production systems: the very old open channel falaj system found in southeast Arabia, and a potential new system based on recycled water. Historically, there were two major agricultural production systems in the Gulf, namely rain-fed rangeland and fully irrigated crop production, regulated not by natural systems, but by human extraction of groundwater.

1. Rain-fed systems

Until recently, most of the area was desert rangeland with livestock populations managed according to systems that maintained productivity and diversity. The only water input to the ecosystem was in the form of rainfall (so-called green water). Average annual rainfall over the entire region is less than 70mm, while potential annual evapotranspiration rates commonly exceed 2500mm. Groundwater recharge rates are therefore very low.

2. Irrigated systems

The other production system, full irrigation, is the only way to reliably produce crops, and this requires large amounts of freshwater (so-called blue water) to meet the high evapotranspiration rate. Irrigation is by far the largest consumer of water in the Gulf. Before the exploitation of deep groundwater began, irrigation was by necessity limited to areas where good quality groundwater was easily accessible, such as from springs.

The Al Ain/Buraimi area in UAE and Oman is an example of an area that is dependent on irrigation, as is the Hofuf area in Saudi Arabia. Another area with significant irrigated agriculture is the Batinah coast on the Gulf of Oman, which historically was irrigated using water lifted from dug wells using animal power, or by aflaj (singular: falaj). The shallow aquifer underlying the Batinah is recharged by runoff from the adjacent Hajar mountains where rainfall is significantly higher.

Whereas groundwater was previously extracted using animal power at a long-term sustainable rate, it is now easily exploited using motor driven pumps. The irrigated area and the rate of groundwater exploitation has consequently increased significantly in the last three decades and has exceeded the recharge rate, resulting in saline water intrusion as the upper freshwater layers are depleted. This is particularly

a problem near the coast, and it has rendered formerly productive land unusable because the salt concentration of well water now approaches that of sea water. Now not even date palms can grow in some areas of Al-Batinah.
Along with the increase in the human population has come a large increase in the livestock population. Figure 3 shows the population of the major livestock species, namely camels, cattle, goats and sheep. The large increase in livestock has been made possible by increased irrigated forage production based on the use of non renewable groundwater. The livestock are also often permitted to graze the rangelands, with the inevitable consequence of overgrazing and degradation. Biomass production has decreased, often accompanied by an increase in unpalatable species. The rangelands often simply cannot sustainably support the increased numbers of livestock made possible by the large-scale introduction of irrigated forage crops such as Rhodes grass (Chloris gayana) and alfalfa (medicago sativa). Destruction of the rangeland will continue as long as it is overexploited.

Figure 3: Livestock Population of Gulf Countries.

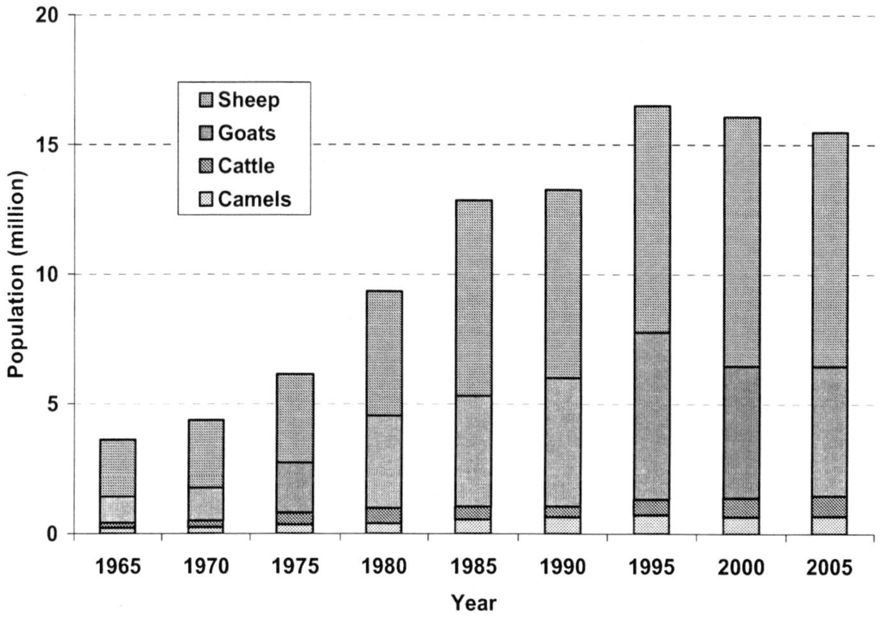

Until recently, it was often the policy to maximize agricultural production and to strive for self-sufficiency in selected agricultural commodities, even though resources such as water could not sustain such production for long without being depleted. This policy was often supported by subsidies. When water, energy and other inputs are cheap due to subsidies, they tend not to be used efficiently. Crops which, from resource and environmental viewpoints are not suitable, may be grown profitably but they may not be wise choices. An example is the widespread production of fully irrigated perennial forages. Large amounts of non-renewable water are consumed, and livestock populations rise beyond a sustainable level. Extension of research to the end users is particularly weak due to the lack of suitably trained local people who could influence agricultural practices.

Much of the management of agricultural production is in the hands of producers or business interests, and they may have a short term outlook that, while maximizing profit, causes longer term environmental degradation and resource depletion. Expatriates working under short term renewable contracts are often managing field operations and have little incentive to conserve. In addition, laws governing agricultural production are sometimes not adequately enforced.

The past decades have seen a profound break between the traditional and mostly sustainable use of water for human use, both directly and for food production, and the current situation in which water resources are used at far greater rates than provided by natural processes. On the other hand, the rangeland ecosystem continues to deteriorate, and it will require significant change to halt and reverse this process.

Case studies of the Integrated Management of Water Resources, Ecosystems, Human Well-being and Ecosystem Services in the Gulf region

Case Study 1: Aflaj in Oman as an example of a Fully Irrigated Production System

Aflaj village systems can be found in many areas of the Middle East and other regions, where they are known by different names, including qanat in Iran, kariz in Afghanistan, and foggara in N. Africa. In the Gulf, they are found only in Oman and the adjacent mountainous parts of UAE, such as Fujairah.

The aflaj of Oman continue to be an example of sustainable water use in a fully irrigated village production system. Many have been in existence for centuries or longer, but are now under intense pressure from competing economic activities. The irrigated area in Oman is about 60,000ha, and about half of this area receives water from aflaj systems. Irrigation accounts for 80% to 90% of Oman's entire freshwater consumption. There has been a large expansion in the area irrigated with pressurized water from wells equipped with powered pumps. These pressurized systems can be sprinkler or bubbler as well as surface irrigation, and are referred to as "modern" irrigation systems. Aflaj are surface irrigation systems, and are often referred to as "traditional" systems. There is a widely held belief in Oman, the Gulf, and elsewhere that modern irrigation is more efficient than traditional irrigation. The major problem facing many aflaj settlements is that there are better economic opportunities elsewhere, and so there is a migration of people out of them, particularly the young. Access to modern media and telecommunications exacerbate this trend. Without the input of labor required to maintain aflaj, they are in danger of falling into disrepair and ceasing to be viable entities. The loss of aflaj systems to this process would end a long historical and cultural tradition.

The main characteristic of aflaj systems is a channel through which water is conveyed from a source such as a spring or a water table at higher elevation. Unless the falaj system taps seasonal flows, there is usually flow throughout the year although the rate may vary. Where the water source is groundwater, it is tapped by horizontal underground galleries that intersect the water table and through which water flows under gravity. Once the gallery intercepts the ground surface it flows through an open channel to the village. Once in the village, the water is used to supply the mosque and for domestic purposes before its use in irrigation.

The irrigated area usually consists of a "core" of mostly perennial crops, commonly date palm, surrounded by "awabi" land that is seasonally irrigated, usually in winter when there is more flow than required by the core crops. Wheat and vegetables are often produced on awabi land. Aflaj villages are managed on a communal basis, with the members receiving a time share of the entire irrigation flow in a regular rotation period, usually of one or two weeks. Norman et al. conducted an on-farm case study of an Omani falaj village to determine how efficiently it used water. The system is located in the foothills of the Hajar mountains, and receives water conveyed in an open channel from a spring at higher elevation. The system is still managed in the traditional way. The time shares vary in length, and the beginning and end times of each timeshare are determined by sun dial during the day and observations of stars at night. Complex rules have evolved over time to take into account day length differences between winter and summer, and to more equitably alternate daytime and nighttime allocations.

Norman et al. (1998) made measurements of water flow to selected individual plots using flumes that, along with flow duration, enabled the volume and therefore depth of irrigation to be measured. Measurements of soil characteristics, root zone development, and soil water content were also made. Potential evapotranspiration was estimated using on-site weather measurements. Farmer interviews provided insight into their water management and the issues they faced. The above measurements, combined with a microcomputer based irrigation scheduling program, was used to estimate daily soil water content over time. Core plots growing date palm, and plots growing wheat in awabi land were monitored.

Results showed that the calculated water demand/supply ratios were mostly indicative of management that did not deliver excess water to the plots. Only in the few days following a significant rainfall event was excess water applied, both because of the direct effect of the rainfall and because higher flow rates occur. Without a way to store significant amounts of water, it is applied in excess during these relatively infrequent times. In dry years with low falaj flow, soil water content often falls below the desired limit, but not low enough to cause significant crop loss. Although the water has no direct cost, when the value of system maintenance and labor is accounted for, the cost of the water is relatively expensive, more so than the cost of water to farmers relying only on motor driven pumps to pressurize water. When water is cheap, it is often used more wastefully. In this case, it was apparent that farmers understood how to manage water carefully and, although it was a traditional surface irrigation system, it was managed better than many modern systems where simply turning on a pump powered by cheap energy is all that is required to provide water. Thus, management is critical. A well managed traditional system can be more efficient than a poorly managed modern system.

The challenge is to preserve traditional indigenous knowledge of water management while providing a livelihood that is competitive with alternatives in other sectors of the economy. Assistance with technical factors such as increasing water storage capabilities as well as with marketing and crop selection will help. Perhaps the introduction of specialized high value niche crops, such as culinary and medicinal herbs will also help. Extra income from limited eco-tourism is another possibility. Appropriate methods and policies must be developed if aflaj are to succeed in the very different world from that in which they were created.

Case Study 2: A New System with Potential to Reduce Grazing Pressure on the Rangeland.

Most of the Gulf countries have mechanisms to recapture some municipal and domestic water and treat it, usually to a tertiary level. This "recycled" water is often used for irrigation of urban landscaping. A fundamental issue with using recycled water for irrigation is that demand varies during the year according to evapotranspiration. Plants, landscape or otherwise, require much less water in the winter than they do in the summer. Figure 4 shows estimated daily potential evapotranspiration for an 18 month period in the Emirate of Sharjah. It can be seen that the minimum values in winter are in the range of 2 to 3mm per day, while the maximum values in summer are typically four times higher, at 10 to 11mm per day and sometimes more. To fully irrigate a fixed area of perennial vegetation, such as landscaping, will thus require a flow of recycled water that is four times larger in summer than it is in winter. However, the production of recycled water is relatively constant during the year due to the capacity of the treatment plant. Without the ability to store the recycled water there is an excess of it in winter, and this is sometimes simply discharged back into the sea.

Figure 4: Daily potential evapotranspiration from February 1999 to August 2000, estimated with the Penman-Monteith method, using weather data from an inland location in the Emirate of Sharjah, UAE.

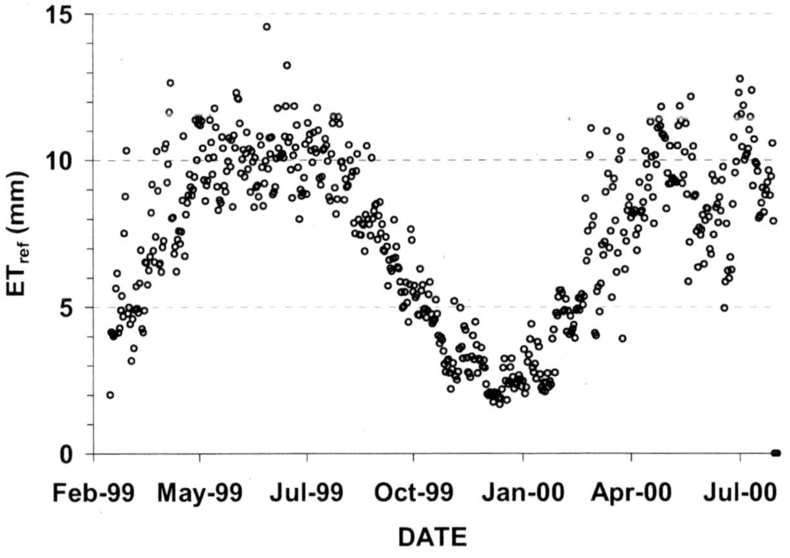

In the Gulf, there is some reluctance to use recycled water for products grown for human consumption, but it is generally considered acceptable for such water to be used for crops that animals will consume. There is thus a potential to use seasonal excess recycled water for forage production. However, if the forage crop is to be fully irrigated, it will have to be an annual forage that is only grown in winter. An advantage of perennial forage crops that makes them attractive is that once they are established they are relatively easy to maintain as they do not require replanting every season. A planting may last a number of years before its productivity drops enough to warrant replanting. The problem with using excess recycled water to irrigate the most widely grown perennial forage crops is that it is not available during the summer because it is all being used for other irrigation purposes, such as landscaping, and without irrigation the crop will die.

A potential solution is to use forage crops that are sufficiently drought tolerant to survive the summer without irrigation yet can yield well when water is available. Native species may be good candidates (Peacock et al., 2003). With such crops, there is the potential to create a very flexible source of forage that could use water from multiple sources, including recycled water, rain, conventional groundwater and brackish groundwater (as long as the salt balance is maintained within the crops tolerance range).

Example of Dubai, UAE

Dubai, like most Gulf cities, depends on desalinated water. Recognizing the potential of treated wastewater, Dubai has been expanding its wastewater collection network and has one of the highest rates of treatment in the region. Table 1 shows the amount of water supplied, lost in distribution, and recycled (wastewater treated to a tertiary level) in 1999. In some cities in the Gulf, the water table has risen in recent years due to excess urban irrigation, drainage and leakage over many years, which may pose a threat to buildings.

Table 1: Water supply, distribution losses and wastewater treatment in 1999 in Dubai, UAE (million m^3/year). Source: Dubai Electricity and Water Authority.

Groundwater	Desalinated	Total	Distribution Losses	Treated Wastewater
10.0	161.0	170.9	23.4	65.7

In this case study, an assumption is made that wastewater is collected and treated at the same rate each month, ie. 5.48 million m^3. With this assumption, the maximum area that can be irrigated in summer is less than it is in winter because of the different crop water requirements. If peak summer crop water use is 10.9mm/day, and if irrigation efficiency is 80%, then the available flow of recycled water of 5.48 million m^3 will be sufficient to fully irrigate 1290ha. Conversely, if the minimum winter crop water use is 3.45mm/day, and assuming the same irrigation efficiency but with a contribution of 20mm from rainfall, then the flow would be sufficient to irrigate 4900ha. If full irrigation is desired throughout the year, then the irrigated area would have to be limited to the summer value of 1290ha. The consequence of this is that, while all the recycled water is used during the peak period, there would be an excess at other times of the year. The excess would amount to about 27 million m^3 per year, or about 41% of the total annual amount of recycled water, with most of this being available in the cooler winter months. This water could be used to irrigate a drought tolerant perennial forage crop. Total annual yield would not be as high as it would be under full irrigation, but the partial irrigation would enable significant production using a water resource that might otherwise not be used productively. If this production was used to keep livestock off the rangelands, particularly during critical times, then perhaps some natural restoration could occur. However, if the production simply resulted in yet more livestock, it would have a negative impact. Development and enforcement of grazing restrictions designed to restore the rangeland would be essential.

Another advantage to this system would be that it would make effective use of any rainfall, and may enable other water sources, such as brackish groundwater, to be used as supplements in a management plan that leached salt buildup before it created problems. Thus, some irrigation with saline water may be possible in the summer as long as salt accumulation could be leached in the winter with recycled water and rainwater before it could build up sufficiently to damage the crop or cause soil salinization.

To this end, a study was conducted in Sharjah Emirate in 2000 to determine yield of a drought resistant forage crop under partial irrigation. The crop selected was Cenchrus ciliaris, which occurs naturally in the area and which therefore must be sufficiently drought tolerant. Peacock et al. (2003) provide details of the collection and identification of this species. A line-source irrigation system was used to deliver an irrigation gradient to plots of Cenchrus ciliaris, and also to Rhodes grass (Chloris gayana), the most commonly grown fully irrigated forage crop in the area.

The irrigation gradients range from full irrigation close to the sprinkler line to zero beyond the range of the sprinklers. Following a cut under an irrigation gradient, full irrigation was resumed until the next cut. This was done to promote development of the root system and to ensure the survival of Chloris gayana. The plots were first established in winter and were subject to an irrigation gradient until the first cut. It became apparent that Chloris gayana could not be successfully established under this irrigation regime due to the resulting soil water stress.

Following the first cut, the plots received full irrigation and we were able to establish Chloris gayana under these conditions. Cuts 3 and 4 received an irrigation gradient and full irrigation respectively. Yield samples were taken in 2m increments perpendicular to the sprinkler line. Irrigation water as a function of distance from the line was measured with catch cans. It can be seen that Cenchrus ciliaris yields about as well as Chloris gayana.

However, there is no data for Chloris gayana from the first cut because we could not establish it, as noted above. It survived cut 3 under the irrigation gradient, but it is unlikely to have survived if full irrigation had not resumed for the following cut.

Figure 5 shows the relative yield of Cenchrus ciliaris for the two cuts under an irrigation gradient (winter and summer). The maximum irrigation was the amount close to the line source (full irrigation), and the maximum yield was from the appropriate plots, and represents the yield under full irrigation. Further work should be done to determine the yield under zero irrigation or very low irrigation, as the relationship shown in Figure 5 may not be accurate at these extreme irrigation levels. However, it can be seen that yield decreases with irrigation, as expected, and reaches a maximum at full irrigation. Because Cenchrus ciliaris is drought tolerant, it will survive with no irrigation, although it is unlikely to produce any significant yield. Any level of irrigation should produce a yield response, and so a forage production system based on a crop such as this can be very flexible.

Of course, the ultimate way to protect the rangeland would be to restore the management practices and livestock populations that prevailed before the rapid economic expansion and population increases caused the degradation. This option would require a drastic decrease in livestock, and a strongly enforced common policy amongst the neighboring countries to control grazing. Imports would have to be increased to even higher levels than currently exist. Although imports represent "virtual" water, there is still a desire to produce traditional products such as livestock.

Figure 5: Irrigation and relative yield of Cenchrus ciliaris for two cuts under an irrigation gradient (winter and summer).

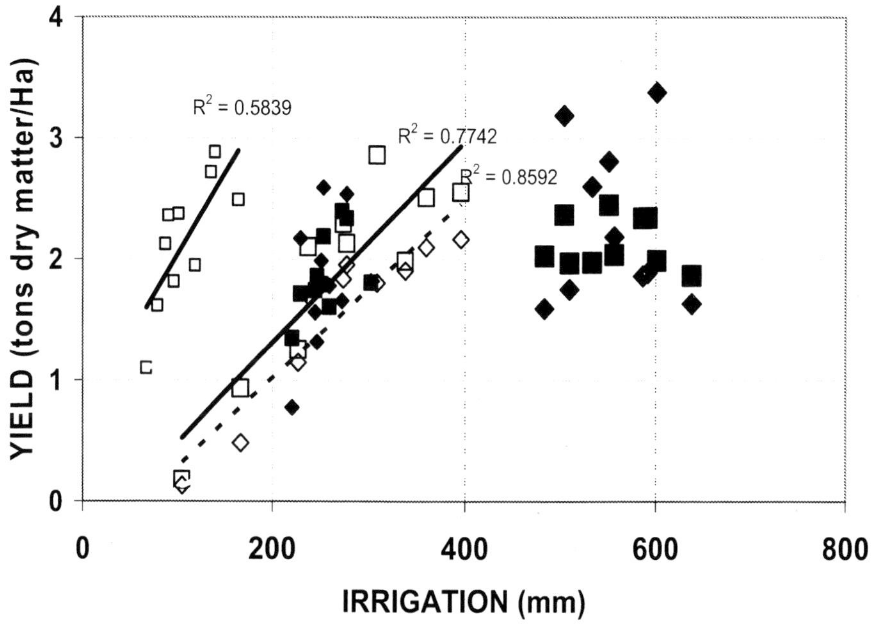

Figures 6 and 7 show hypothetical estimated forage production for Cenchrus ciliaris using the figures in Table 1 for recycled water production in Dubai, the yield versus irrigation relationships of Figures 4 and 5, and representative values for evapotranspiration and rainfall. It can be seen that the maximum total production occurs when about 5000ha receives whatever recycled irrigation water is available after meeting the requirements for full irrigation, such as landscaping, parks etc. Below 5000ha the yield per ha is higher because more water is available per ha, but this is more than offset by the smaller production area. Above 5000ha the yield per ha is lower, and this is not sufficiently compensated by the increased area. While the absolute values used in this case study may vary, there is still likely to be an optimum area of partial irrigation. This analysis could be expanded to include multiple years and incorporate variability in rainfall and projected changes in the volumes of recycled water, as well as the use of other water resources, such as brackish groundwater. The most effective means of delivering the irrigation water would also need to be considered. It is not clear whether such a partially irrigated production system would be economically cost effective in comparison to imports, but it would have some non-monetary benefits that might make it attractive to the Gulf countries with their considerable financial resources.

Figure 6: Hypothetical estimated forage production for Cenchrus ciliaris.

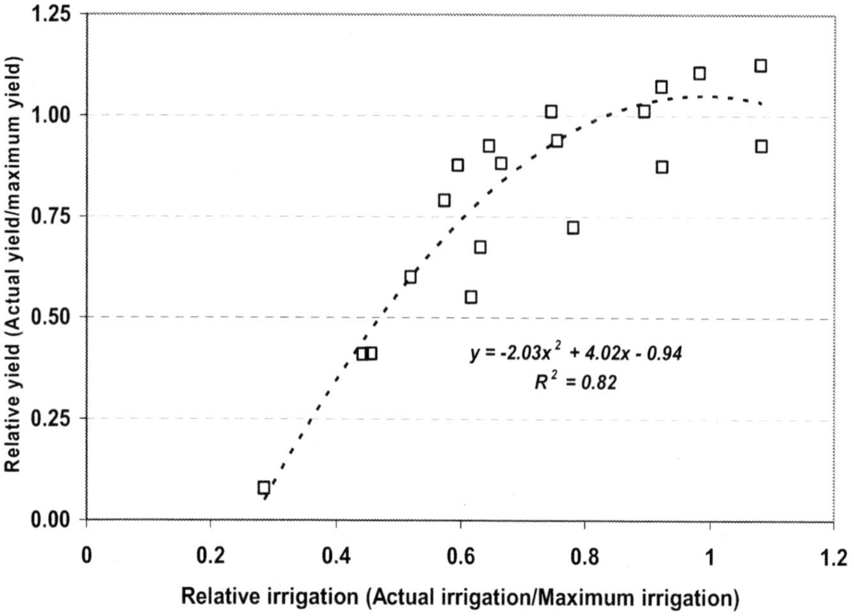

Figure 7: Potential annual production of partially irrigated Cenchrus ciliaris forage using treated wastewater volumes available in Dubai, UAE, as a function of the area irrigated.

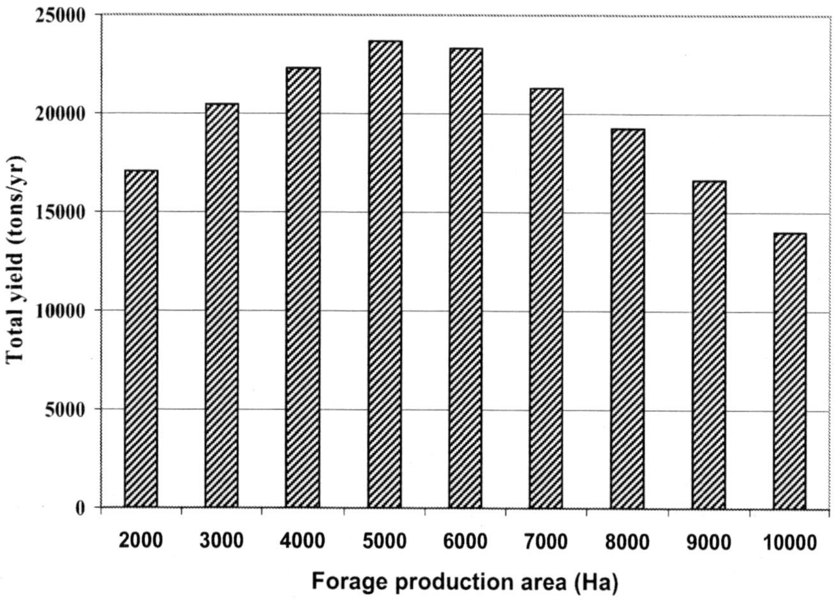

Gaps and Capacity Development: Institutional, Managerial, Human, Technological Capacity

There is great value in the ecosystems of the Gulf in terms of the genetic resources of plants that are adapted to extreme heat and drought. There is also much value in the indigenous knowledge of the people who have lived in these ecosystems without destroying them. Policies that encourage conservation of water and restoration of ecosystems are needed. Water is still relatively cheap, especially considering that it often originates from desalination or non-renewable groundwater. Increased prices do encourage conservation, as can be seen in Dubai where expatriates pay considerably more for municipal water than do local people and have lower per capita water use.

Conserving water and protecting/rehabilitating the environment may require a net decrease in agricultural production and total economic value in the case of rangelands, or recognition and compensation for local products produced in an environmentally sound cost manner in the case of aflaj. Current economic policies have produced non-sustainable systems. There is a cost to sustainable production, and ways must be found to pay for this cost if the end result is not to be destruction of ecosystems and sustainable human settlements. In general, current and projected consumption of food cannot be met sustainably by the region's ecosystems without imports. The challenge is how to adapt to this situation without doing irreversible damage to the fragile systems which used to support the population but which now cannot.

Policies often do not encourage or fund innovation and conservation. While there are some good universities and research institutions in the GCC, investment in agricultural and environmental research is often not a high priority item in terms of funding and human resource development. Much of the skilled personnel, from researchers to technicians and other support people, are expatriates with no long term future or stake in the countries.

References

Al-Rashed, M.F. and M.M. Sherif. (2000) "Water resources in the GCC countries: an overview", *Water Resources Management* 14(1): 59-75.

Böer, B. (1997) "An introduction to the climate of the United Arab Emirates", *Journal of Arid Environments* 35(1): 3-16.

Norman, W.R., W.H. Shayya, A.S. Al-Ghafri and I.R. McCann. (1997) "Aflaj irrigaton and on-farm water management in northern Oman", *Irrigation and Drainage Systems* 12(1): 35-48.

Peacock, J.M, M.E. Ferguson, G.A. Alhadrami, I.R. McCann, A. Al Hajoj, A. Saleh and R. Karnik. (2003) "Conservation through utilization: a case study of the indigenous forage grasses of the Arabian Peninsula", *Journal of Arid Environments* 54(1): 15-28.

Zubari, W.K. (1998) "Towards the establishment of a total water management and reuse program in the GCC countries", *Desalination* 120(1998) 3: 14.

The Lake Victoria Region

Ndalahwa F. Madulu

Institute of Resource Assessment, University of Dar es Salaam, Tanzania

Introduction

"Our countries, the Nile Basin Countries, are witnessing a revolutionary development era to face the challenges of water scarcity in the 21st century – the main goal is to adopt an integrated water resources management program (H.E. Mahmoud Abu Zeid)"

Water is an important commodity for human beings, animal and plant life. It is a vital component of sustainable development and poverty alleviation at all levels. Provision of water services is one of the most important prerequisites for improving quality of life for people (URT, 2002), which is the long-term goal of almost all East African countries. Approximately 80 percent of the population of East Africa live in rural areas, and only about 37 percent of the rural population has access to *'safe water'*. Lack of access to reliable and safe water is a major hindrance to the attainment of the MDGs in the East African context.

Although effective and sustainable management of water resources is essential, water resource management systems in East Africa seem to ignore the approaches that would lead to sustainable water resource management. In many parts of East Africa, freshwater resources are used up to the marginal limits and no sustainable management and operational systems have been put in place to ensure sustainability (Zaba and Madulu, 1998). Moreover, surface water is fouled with pollutants from industries, municipalities, and agriculture. The lack and poor quality of water is associated with many incidences of water-borne diseases such as typhoid, diarrhea, and cholera, which have remained major problems in many regions of Tanzania.

The purpose of this paper is to discuss issues related to sustainable management of water resources in the Lake Victoria basin. Based on the understanding that the lake is a shared water resource between Tanzania, Kenya and Uganda, joint efforts are being undertaken to address major environmental problems in the lake basin in an integrated manner. An attempt is also being made to identify the conflicts that emerge due to various competing interests and demands from water users. It is argued that adoption of *Integrated Water Resource Management (IWRM), Sector-Wide Approaches (SWAPs), River Basin Management Approach (RBM)*, and joint planning and management of the lake resources, could reduce the occurrence of major environmental problems in the basin. Collaborative efforts are being made in the areas of policy decisions, resource utilization, and exchange of expertise and information related to the lake resources.

Water Resources of Lake Victoria

Lake Victoria is the world's second largest freshwater lake, and the largest in Africa, with a surface area of 68,800 km^2. It provides habitat to a myriad of fish species and a variety of other species and biodiversity richness (Bootsma and Hecky, 2003). The lake is shared between Kenya (6%), Tanzania (51%) and Uganda (43%). The lake catchment area covers 193,000 km^2 with Tanzania occupying 44 percent, Kenya 22 percent, Uganda 16 percent, Burundi 7 percent and Rwanda 11 percent. Figure 1 shows the catchment area and main drainage features of the Lake Victoria basin.

Figure 1: The Lake Victoria Catchment Area and Its Main Drainage Features

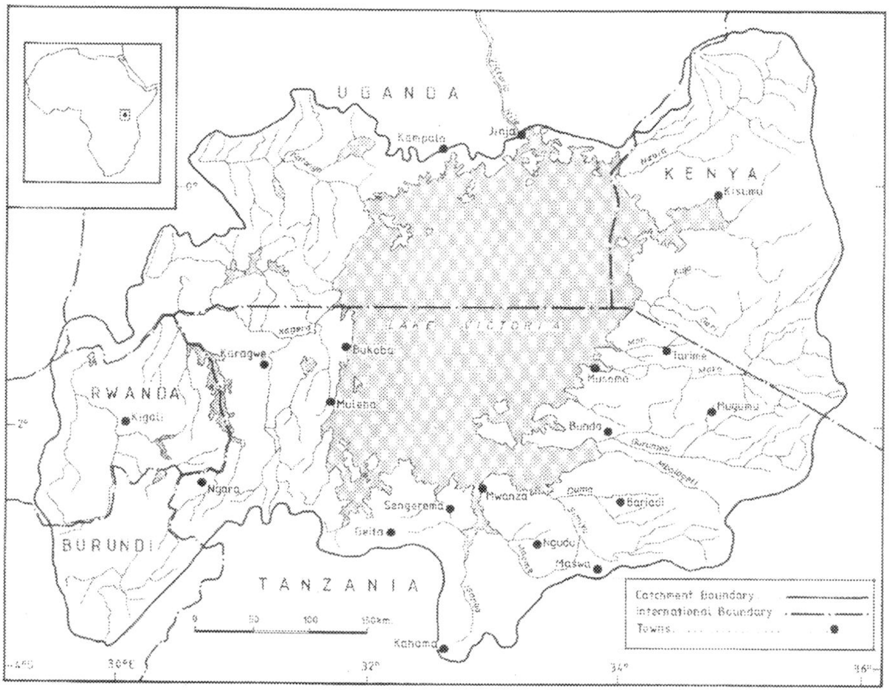

Despite the size of the lake, much of the surrounding population suffers from lack of access to water supplies. Recent estimates from the Ministry of Water and Livestock Development in Tanzania indicate that only 50 percent of the rural population and 69 percent of the urban population in Tanzania have access to reliable water supplies. Similarly, less than 60 percent of the Ugandan population have access to safe water supply or sanitation services, and in both cases, many rural facilities no longer function (Mutayoba, 2002; Zaba and Madulu, 1998). In rural areas of Kenya, such as those close to Lake Victoria, only 31 percent of the population had access to safe water supplies in the year 2000. Although basin level data on water supply were not available for this study, Lake Victoria and its catchment support about 30 million people (Machiwa, 2003; UN-Habitat, 2004). This constitutes about one third of the total population of Kenya, Tanzania and Uganda. Furthermore, because Lake Victoria is part of the Nile River Basin, its water resources are therefore of great interest to all of the riparian nations. Table 1 compares the characteristics of Lake Victoria to other large freshwater lakes in the world.

About 3 million people earn their living directly or indirectly from the fishing industry of Lake Victoria in the three countries (Hoza, 2003). Lake-wide fish production is estimated at between 400 – 500 metric tons with Tanzania landing 40%, Kenya 35% and Uganda 25%. The landed value of this catch is between USD 300 – 400 million annually. The income generated from these activities has multiplier effects on the local economy. While fishing is an essential source of protein to the local population, it also provides large amounts of foreign exchange revenues to the governments in terms of fish exports. Similarly, it provides income to the local people who are directly and indirectly engaged in fishing-related businesses.

Table 1: Position and characteristics of Lake Victoria as compared to other world largest freshwater lakes

Characteristics	World Largest Lakes by Rank								Rank of Lake Victoria
	Superior	***Victoria***	***Huron***	***Michigan***	***Tanganyika***	***Malawi***	***Erie***	***Ontario***	
Surface Area (km2)	82,100	68,800	59,800	57,750	32,600	29,500	25,800	19,000	2
Drainage Area (km2)	128,000	195,000	134,000	118,100	220,000	100,500	61,000	64,000	2
Volume (km3)	12,230	2,760	3,537	4,920	18,900	7,775	483	1,637	6
Altitude (m amsl)	183	1,134	177	177	774	474	174	75	1
Maximum Depth (m)	407	79	229	282	1,470	700	64	245	7
Mean Depth (m)	149	40	59	85	580	264	19	86	7
River Inflow (km3/yr)*	50	20a	165	36	14b	29c	196	229	7
River Outflow (km3/yr)	71	20a	170	47	2.7b	12c	196	230	6
Rainfall (km3/yr)	65	100a	51	47	29b	39c	24	17	1
Evaporation (km3/yr)	48	100a	40	42	50d	57e	24	13	1
Residence Time (years)	107	23	16.4	59	440	114	2.2	6.7	5
Flushing Time (years)	172	138	21	105	7,000	648	2.5	7.1	4

Source: Bootsma and Hecky, 2003:5

In addition to providing fishing opportunities and food (fish), the lake water is used to generate hydro-electric power, transport (and communication), tourism, water for domestic agricultural and industrial use, and recreation. Given these positive local, national, and regional benefits, conservation of natural resources in the lake basin becomes vital and necessary to the survival of the local communities and biodiversity conservation.

Drivers of Change in the Lake Victoria Basin

It has been observed that most of the environmental changes currently occurring in the Lake Victoria Basin are induced by human activities. These activities include agricultural expansion, deforestation for provision of essential needs for the people's livelihoods (food, shelter, clothing, etc.), mining operations, fish processing, and rapid urbanization leading to increased charcoal demands (Yanda et al., 2001; Liwenga et al., 2005).

Demographic factors are particularly important in driving these changes in the Lake Victoria ecosystem and its management. Table 2 summarizes the continual population growth in four regions located in the Lake Victoria basin in Tanzania, as observed from the 1967, 1978, 1988 and 2002 population censuses. By the time of the 2002 census, the lake basin regions accommodated over 9 million people, representing about 27 percent of the Tanzanian population (URT, 2003). Both in the 1988 and 2002 population censuses, Mwanza and Shinyanga Regions were ranked first and second in the country in terms of population size. Kagera Region was ranked sixth and fifth in the two censuses, respectively.

Over 70% of the population in the Lake Victoria Basin is engaged in agricultural production, mostly as small-scale farmers. Table 3 shows the proportion of agricultural land in the Lake Victoria basin which is under cultivation.

Poverty in the Lake Victoria basin is remarkably high. In Kenya, for example, about 50%, 63% and 59%, of the population in the Rift Valley, Nyanza and Western Provinces, respectively, are classified as being poor. Poverty is often evident among the low income groups due to various factors including poor skills, lack of capital, poor infrastructure, diseases, and lack of education.

Population growth and poverty are reflected in many mutually reinforcing negative social and environmental trends. On the one hand, there is rising demand for environmental resources like water, fuelwood and arable land. On the other hand, over-fishing, removal of forest cover, land degradation and deforestation are leading to reduced water quality, decline of endemic fish species, loss of biodiversity and declining agricultural productivity (Bootsma and Hecky, 2003; Machiwa, 2003; Odada, 2004). Thus, the degradation of natural resources in the basin further limits the basin's capacity to support current and future populations (Madulu, 2005). The resulting insufficient food supplies and lack of access to safe water lead to waterborne diseases, malnutrition and increased infant and child mortality rates. These problems are further compounded by poverty and poor access to health facilities (UN-Habitat, 2004).

Table 2: Population Size and Growth rates of the Tanzanian Lake Victoria Regions (1967-2002)

Regions	Census Population Data				Intercensal Growth Rate	
	1967	**1978**	**1988**	**2002**	**1978-88**	**1988-02**
Shinyanga	899,468	1,323,535	1,763,800	2,805,580	2.9	3.3
Kagera	658,712	1,009,767	1,313,594	2,033,888	2.7	3.1
Mwanza	1,055,883	1,443,379	1,876,635	2,942,148	2.6	3.2
Mara	544,125	723,827	946,418	1,368,602	2.9	2.5
Total	**3,158,188**	**4,500,508**	**5,900,447**	**9,150,218**	**2.7**	**3.1**

Source: (URT, 2003:2)

Table 3: Agricultural Characteristics of Lake Victoria Basin

Country	Catchment Land Area (1,000ha)			
	Cultivated	**Non Cultivated**	**Total**	**Percent Cultivated**
Kenya	1,470	3,400	4,870	30.2
Uganda	1,400	2,100	3,500	40.0
Tanzania	1,500	5,540	7,040	21.3
Rwanda	930	1,130	2,060	45.1
Burundi	670	640	1,310	51.1
Total	**5,970**	**12,810**	**18,780**	**31.8**

Source: (Odada, 2004:18)

Ong'ang'a et al. (2001) looked at the basin's ecological trend in the light of cultural practices in land management that enhance conservation. He concluded that most of the past cultural practices implemented by various ethnic groups to ensure sustainable management of natural resources in the past have been abandoned and this has resulted in massive degradation of the ecosystem and the lake basin as a whole.

Unsustainable utilization of the major wetland areas through agricultural activities and livestock overgrazing has greatly compromised their buffering capacity. Soil erosion, caused by changes in agricultural practices, is one of the main processes of land degradation in the Lake Victoria Basin. Yanda et al. (2001) argued that this process is assumed to lead to depletion of soil nutrients in agricultural land, degradation of wetlands through siltation and sediment deposition into the lake. These processes therefore affect the fish habitat and ecology of the lake.

The lake ecosystem is further affected by the inflow of polluted runoffs generated by new land-use practices. Agro-chemical residues (herbicides and pesticides and industrial fertilizers), and to a limited extent, heavy metals resulting from gold mining operations contribute to eutrophication of the lake (Yanda et al., 2001; UN-Habitat, 2004). The increasing population density and urbanization in the basin are additional factors contributing to the creation of pollution hot-spots leading to localized degradation of water quality in the lake from human waste, urban runoff, domestic effluent, agricultural and industrial discharges (LVEMP, 2004; Machiwa, 2003). According to UN-Habitat (2004), the collection of nutrients (phosphorus and nitrogen) emanating from human and industrial wastes have increased five-fold the algae growth in the lake since 1960s, causing de-oxygenation of the water which threatens the survival of deep water-fish species.

Challenges to Improve Water and Ecosystem Management in the Lake Victoria Basin

The biggest challenge in the Lake Victoria basin is to expand water and sanitation services in rural and urban areas in order to reduce the pollution impacts and incidences of water-borne diseases. According to the Tanzanian National Strategy for Growth and Reduction of Poverty (NSGRP)[2], there is a close link between water supply and waterborne diseases such as cholera, bilharzias, malaria and typhoid. Conversely, in some cases, pollution of the lake waters has been the most obvious result from poverty alleviation activities (Yanda et al., 2001; Machiwa, 2003; Liwenga et al., 2005). While dealing with poverty reduction on the one hand, on the other hand the instituted poverty reduction strategies have negative ecological and sustainability impacts.

It is clear that a new integrated approach to water resources management is needed to ensure sustainable management of the Lake Victoria basin. Considering both temporal and spatial water requirements to maintain the health and viability of ecosystems, there is need to institute the *"Polluter Pays"* principle and adopt Integrated Water Resource Management (IWRM) approaches which are cross-disciplinary in nature and are undertaken in order to manage water in an equitable manner without compromising the sustainability of the ecosystems (GWP, 2000; Duda and El-Ashry, 2000; Lundqvist et al., 1985). Such approaches recognize the

[2] The National Strategy for Growth and Reduction of Poverty is widely referred to as MKUKUTA (Mkakati wa Kukuza Uchumi na Kuondoa Umasikini Tanzania)

important role that water ecosystems play in the national economy, and take full account of the linkages between land use and water management issues. IWRM is based on recognition of the regular interactions, uses and interests of various stakeholders that converge around the common resource. In the case of Lake Victoria, these include agriculture, industry, fishing, wildlife, domestic water supply, hydropower generation, recreation, and the environment (Mutayoba, 2002). Collaborative efforts are being made among East African countries to establish institutional frameworks that are capable of effectively coordinating resource use and conservation in the lake basin.

Progress in Integrated Water and Ecosystem Management and Policies for Lake Victoria

To ensure effective IWRM strategies in the Lake Victoria basin, the East African countries agreed through the Lake Victoria Development Programme (LVDP) that:

- IWRM processes should be established along hydrological boundaries,
- All relevant stakeholders should be involved in the IWRM processes for the basin,
- Ecosystem service valuation should be included in the decision-making processes,
- Countries should monitor and report on the success of IWRM processes.

The LVDP was established in 2001 to coordinate the various interventions on the lake and its basin; and serves as a centre for promotion of investments and information sharing among the various stakeholders. To support these agreements, the East African countries established in 2003 the Protocol on Sustainable Development of Lake Victoria and its Basin, which gives the legal backing for all the intended interventions in the Lake Basin. The implementation of the agreements between the members of the East African Community is a major challenge that would have an impact on Lake Victoria basin's ecosystem and its resources.

Successful examples of IWRM that are worthy noting include the joint efforts among the East African countries and deliberate community involvement that have led to reduction in the prevalence of *water hyacinth* in Lake Victoria in recent years. The invasion of *water hyacinth* in the Lake Victoria was previously seen as a major environmental problem in the lake especially during the late 1990s and early 2000.

There are a number of significant initiatives to support IWRM currently taking place on Lake Victoria:

- The Lake Victoria Environmental Management Project (LVEMP) is a regional and comprehensive environmental development programme which evolved through a process guided by the Tripartite Agreement of 1994 between Kenya, Tanzania and Uganda. The LVEMP aim was to restore a healthy, varied lake ecosystem that is inherently stable and can support, in a sustainable way, the many human activities in the catchment and the lake itself (LVEMP, 2004, Machiwa, 2003).

- The Lake Victoria Region Local Authorities Cooperation (LVRLAC) to promote the sustainable development and management of the Lake Victoria resources for the benefit of the basin's urban dwellers. LVRLAC focuses on poverty alleviation, improvement of the environment and development of the lake region (UN-Habitat, 2004).

- The East African Community's Organization for the Management of Lake Victoria resources (ECOVIC), whose objectives include promoting and coordinating economically viable natural resource utilizations that maintain biodiversity and safeguard the aquatic ecosystem; gathering and dissemination of information on key environmental and socio-economic issues within the lake region; research, lobbying and advocacy for favourable policies and community participation on issues affecting Lake Victoria and its basin; promoting food security in the basin through sustainable agriculture and fishing, promoting proper sanitation and hygienic conditions along the fish landing beaches and on the lake's islands; and reducing the spread of HIV/AIDS and its impact on the communities in the lake region. (UN-Habitat, 2004; Hoza, 2003).

- The Lake Victoria Region Water and Sanitation Initiative (LVRWSI), which aims to improve water and sanitation facilities in the secondary towns in the Lake Victoria basin by combining physical investments in infrastructure provision, with targeted capacity-building (UN-Habitat, 2004).

- The Lake Victoria basin is also part of the Nile basin. The broader Nile Basin Initiative (NBI) is a regional institution, established in 1999 after long negotiations, aimed at equitable sharing of the Nile River waters and benefits among the riparian countries (UNCSD, 2005).

In the near future, a Lake Victoria Water Office will be opened to supervise water resource use and distribution among various stakeholders in Tanzania. Management of water resources in Tanzania is already vested in the Basin Authorities (URT, 2002). In support of this approach, there are other efforts that are being implemented in the basin. These include the Mara and Kagera sub-basins. The Mara River runs across the border into Kenya, and the Kagera river flows through Rwanda, Tanzania and Uganda. These are internationally shared rivers and hence, deserve a cross-border management system. A joint management approach in the form of the Demonstrative Framework was agreed by the East African Community in 2000 with the purpose of realizing joint management and equitable utilisation of lake resources.

At the community level, there have been notable successes in water and environmental management around Lake Victoria which should be replicated across the lake basin. In the rural areas of Mwanza Region, local communities usually organize themselves to construct charco dams (*bwawa*) for their livestock (Madulu and Zaba, 1998). The local communities normally enact and effect by-laws to limit human activities that pollute the water sources. In Kibuyi village in Tarime District, local communities participated in the construction of community toilets and bathrooms along the beach to minimize pollution (Liwenga et al., 2005). Under the HESAWA programme, improvement, management, and operation and maintenance (O&M) of water sources were the responsibilities of the respective local communities, hence, ensuring community ownership and accountability (Zaba and Madulu, 1998).

Capacity Development Needs and Gaps

A previous assessment of policy needs by Odada et al. (2004) has already recommended the establishment of fishing quotas and quotas for use of the lake by processing industries, as well as generalised civic education and awareness programmes. With regards to destructive fishing practices, these authors recommended strengthening monitoring and enforcement of restrictions and rule of law, providion of civic education and awareness, empowerment and involvement of more communities in management, imposition of size restrictions on fish-processing factories, and provision of credit to artisan fishers. To further address pollution issues, Odada et al.(2004) recommended strategies including the accreditation of analytical laboratories for standards enforcement, liberalization of waste disposal activities to involve the private sector and communities in revising regulations in urban planning to take into account environmental issues and improve monitoring and enforcement of agreed laws and regulations to improve natural resource management.

In order to implement the various policies identified above, there is a need for capacity building at various levels from the Lake Victoria basin level to the regions, districts and village levels. Increased capacity will enable improved supervision and sustainable management of resources in the basin. Odada et al. (2004) also recommended capacity building for the legal and economic empowerment of institutions to facilitate and enforce compliance with international conventions and to increase the effectiveness of the National Environmental Protection authorities.

In another study on capacity needs in order to meet the MDGs and provide water to the cities on the lake, UN-Habitat (2004) identified areas for capacity building to include managerial and financial skills at the regional (sub-national) level, setting up a modern billing and revenue collection system in order to improve revenue collection at local levels, and provision of hardware and software. Capacity building is also needed in the field of integrated physical planning for infrastructure development. Similarly, capacity building is needed for local authority officials in the area of solid waste and wastewater management. Other areas include capacity to address health hazards and threats to the lake ecology; pollution control, and water quality monitoring and enforcement of regulations in place.

All of these capacity building areas need to be strengthened by community mobilization and public awareness raising to address issues in a participatory manner (UN-Habitat, 2004). The emphasis should be on a regional collaborative spirit which is under-pinned by the region-wide capacity-building to enrich the region's ability to manage its resources.

Conclusion

It is clear from the above discussion that a new approach to water resources management is needed to ensure sustainability of the Lake Victoria basin. Countries have started to realize the importance of collaborative efforts in managing the water resources and jointly benefiting from their efficient use. The examples of the NBI and the EAC justify the joint efforts that have been made to develop and utilize water resources in the basin. Although there is a consensus on the need for IWRM and the basin approaches for sustainable management, more efforts are needed to strengthen capacity to effectively manage resources in a sustainable manner.

Countries should continue to work together, exchange expertise, and share information and experiences in order to ensure that the resources in the lake basin are sustainably managed for the benefit of all.

References

Bootsma, H.A. and R.E. Hecky. (2003) "A Comparative Introduction to the Biology and Limnology of the African Great Lakes", *Journal of the Great Lakes Research* 29(Supplement 2): 3-18.

Duda, A.M. and M.T. El-Ashry. (2000) "Addressing the Global Water and Environment Crises through Integrated Approaches to the Management of Land, Water and Ecological Resources", *Water International Journal* 25(1): 115-126.

GWP (2000) *Integrated Water Resource Management,* TAC Background papers, Nomer 4, Stockholm, Sweden: Global Water Partnership.

Hoza, R.B. (2003) *Harmonization of fisheries legislation in the East African Community (EAC), Partner States of Kenya, Uganda and Tanzania* available from www.lvemp.org/L_Whats%20new/fisheries_legislation.htm

Liwenga, E.T., R.Y.M. Kangalawe, N.F. Madulu, M.K. Waringo and R. Kabumbuli. (2005) "Livelihood Diversification and Changing Land Use Patterns in the Lake Victoria Basin: An Assessment of Causes and Implications to Local Communities", Phase One Report presented at the VicRes Annual General Meeting, Ngurudoto, Arusha.

Lundqvist, J., U. Lohm, and M. Falkenmark (eds). (1985) *Strategies for River Basin Management-Environmental Integration of Land Water in A River Basin,* Dordrecht, Boston: D. Reidel Publishing Co.

LVEMP (2004) "Lake Victoria Environmental Management Project Brief", Mwanza.

Machiwa, P.K. (2003) "Water Quality Management and Sustainability: The Experience of Lake Victoria Environmental management Project (LVEMP) - Tanzania", *Physics and Chemistry of the Earth* 28(20-27): 1111-1115.

Madulu, N.F. (2005) "Impact of Population Pressure and Poverty Alleviation Strategies on Common Property Resource Availability in Rural Tanzania", *African Journal of Environmental Assessment and Management* 10: 26-49.

Madulu, N.F. and B. Zaba. (1998) "Water Availability and Rural Migrations in *Development* 5(1& 2): 23-38.

Mutayoba, W.N. (2002) *Management of Water Resources in Tanzania Through Basin Management.* Paper Presented at the 3rd WaterNet/Warfsa Symposium on Water Demand Management for Sustainable_Development'. Dar es Salaam, 30-31 October 2002.

Odada, E. O., D.O. Olago, K.A.A. Kulindwa. (2004) "Mitigation of environmental problems in Lake Victoria, East Africa: Causal chain and policy options analyses", *AMBIO* 33(1-2): 13-23.

Ong'ang'a, O. (2001) "Culture and Conservation: Cultural Practices that enhance the Conservation and Management of Natural Resources around Lake Victoria: Ethnic Communities of East Africa, Lake Victoria Basin", RELMA-Regional No. 5.

UNCSD (2005) "Freshwater management: policy options and possible actions to expedite implementation", United Nations Commission on Sustainable Development, Economic and Social Council, Report of the Secretary-General, 9 December 2004 (E/CN.17/2005/2).

UN-Habitat (2004) "Lake Victoria Region Water and Sanitation Initiative, Project Proposal (Revised), Supporting Secondary Urban Centres in the Lake Victoria Region to Achieve the Millennium Development Goals, An Initiative Under Water for African Cities", Programme Phase II United Nations Human Settlements Programme, Nairobi, 11 November 2.

United Republic of Tanzania (URT) (2002) "United Republic of Tanzania National Water Policy", Ministry of Water and Livestock Development, Dodoma.

United Republic of Tanzania (URT) (2003), "Age and Sex Distribution, 2002", Population and Housing Census, Vol. 2 United Republic of Tanzania, Central Census Office, National Bureau of Statistics Dar es Salaam.

Yanda, P.Z., E.K.Shishira, N.F. Madulu, and A.S. Kauzeni. (2001) *Soil Erosion Hazards as a Basis for Identifying Priority Areas for Conservation in Lake Victoria Basin.* Paper Presented at a Workshop Prepared by Lake Victoria Environmental Management Project (LVEMP). Mwanza August 2001.

Zaba, B. and N.F. Madulu. (1998) "A Drop to Drink? Population and Water Resources: Illustrations From Tanzania", in A. Sherbinin and V. Dompka, eds., *Water and Population Dynamics: Case Studies and Population Implications,* Washington: American Association for the advancement of Sciences.

The Hengshui Lake Wetland

Guan Zhuojin[1] *and Liu Zhenjie*[2]

[1]Beijing Hydraulic Research Institute, Beijing Forestry University, China
[2]Management Office for Hengshui Lake, National Natural Conservation, China

Introduction

Over time, Hengshui Lake has provided for human needs in many different ways. The resulting burden on the lake has progressively increased as the surrounding population has continued to develop and grow. The lake now serves many functions for human society, economy and ecology, including provision of water for irrigation of agriculture, serving as a headwater resource for industry, providing scenic tourist spots, supplying safe drinking water, providing ecological restoration and conservation of biological diversity, and so on. Pressures on the lake that are associated with these functions are increasing. According to plans for the South-to-North Water TransferProject in China, the Chinese government proposes to channel water from the rivers of the South to the drier North (Yang and Zehnder, 2005). This plan will cause Hengshui Lake to be transformed into an enormous reservoir. The challenge for management is how to regulate and deal with the relationships between protecting the wetland ecosystem and establishing a rational system for its use; and how to enable peasants, managers and officers to participate in this system. The local population should take part in the restoration of ecological processes in the wetland, particularly to decrease the accumulation of nutrients and to avoid eutrophication. This will enable biodiversity to increase and ensure the sustainability and health of the wetland ecosystem.

Hengshui Lake Wetland Ecosystem

Hengshui Lake is located in the centre of the North China Plain (Figure 1).

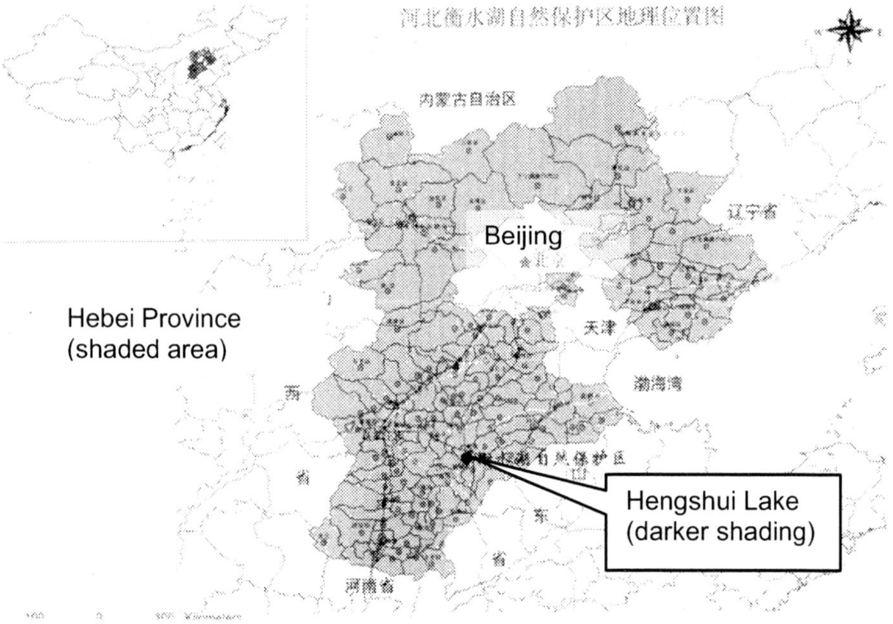

Figure 1: Location of Hengshui Lake, Hebei Province, China.

The lake ecosystem has a semi-arid and dry subhumid climate, controlled by the mainland terrestrial monsoon in the temperate zone. The mean annual climate conditions are air temperature: 13°C, rainfall: 518.9mm, and evaporation: 1,296mm. The annual rainfall is therefore much less than evaporation. Under the current management regime, the water of Hengshui Lake is supplied by the Yellow River, bringing about 80 million tons of water into the 42 km^2 Hengshui Lake wetland. But under the future management regime of the South-North Water Transfer scheme, to come into effect around the year 2010, the water supplied is projected to be 180 million m^3 and to cover an area of 75km^2.

Human Uses of Hengshui Lake Waters

The following human needs for water are supplied by Hengshui Lake waters:

Supply of Safe Drinking Water

Hengshui Lake is to be developed as a source of surface water for drinking. The lake will become an important headwater site for the surrounding urban areas and will bear an important function to supply safe drinking water to millions of local people. The amount of the water to be supplied will be 36 million m^3 annually (perhaps after 2008). Because of this, the quality of water in Hengshui Lake is critical, and its capacity for natural removal of contamination must be protected.

Irrigation for Agriculture

Hengshui city has depended on agriculture for many years. Since 1958, Hengshui Lake has provided water for agricultural irrigation. This practice is continuing in the present day.

Industry

Following the construction of Hengfeng power plant in 1996, Hengshui Lake has provided water for industry as well as for irrigation. The water used for cooling at the power plant since its construction has amounted to 12 million m^3 annually.

Eco-tourism

Recently, in order to develop the regional economy, the local government decided to make ecotourism an important part of its strategy for economic growth. Hengshui Lake will be established as a wetland scenic spot. In recent years, the number of visitors has already increased.

The lake ecosystem is an important area for wildlife as well as for humans. In 2003, a national nature reserve was established in the Hengshui Lake wetland (Urban Programing Institute of Beijing, 2004). This had a positive impact on wildlife and natural resources. The bird population observed in the area increased to more than 296 species. Many ecological experts commented on this phenomenonally dramatic process of ecological recovery (Wang et al., 2001).

Pressures on the Hengshui Lake Ecosystem

Hengshui Lake ecosystem has undergone severe processes of environmental change which are related to physical alteration of the lake habitat, as well as to problems of water quality (eutrophication) and water quantity. These trends are described briefly here:

Physical Alteration

Originally, Hengshui Lake was a natural wetland. It was formed in the shallow basin of the old Yellow River, old Zhang River, old Hutuo River, Fuyang River and their branches. Its area was 75km^2, including two sections of lake, east and west. Only the east lake, with an area of 42.5km^2, retained water throughout the year to a depth of 3-4m. In 1958, an artificial levee was constructed on the original landform of the lake basin in order to transform it into a reservoir. Water could then be impounded and released again (see Figure 2) in order to supply human needs in the Hengshui city region, to the North and Jizhou urban region to the South. This intervention altered the natural hydraulic system of the lake.

Figure 2: Outlet Gate of Hengshui Lake.

Water flows into Hengshui Lake have been reduced gradually, due to the diversion of other rivers upstream and the resulting drying-up of natural rivers. Currently, about 80 million m^3 of water flows into the lake from the Yellow River on an annual basis. However, under the South-to-North Water Transfer Project, water from the Yangzi River will be transferred to the Beijing region. In the new system, the flows into the lake will be greatly increased. The new water storage capacity will be 180 million m^3 and the new reservoir will cover an area of 75km^2. The water quality is expected to be poor. The enlargement of the lake will begin around 2010, in preparation for the new water supplies.

Eutrophication of the Water Body

As human populations have increased in many parts of the world, technological solutions have supported rising living standards and increased food production. Large amounts of chemical fertilizers have been introduced into production systems, leaving high levels of chemical fertilizers and nutrients to remain in the environment.

This leads to the widespread prevalence of eutrophication problems. In the case of Hengshui Lake, the creation of a wetland was regarded as an optimal method to clean the polluted water. Due to this need, increased attention was paid to wetland ecosystems and their ecological processes, usage and protection (Mitsch and Gosselink, 1993; Wang et al., 2002).
According to existing studies (Zhuojin, 2005), the capacity of wetlands for removal of pollution is limited. If the level of pollution exceeds the capacity of the wetland for its removal, this will result in the degeneration, destruction or collapse of the wetland ecosystem. Because of this threat, the use of the wetland should be strictly managed in order to optimize the performance of the wetland in the decontamination of polluted water.

Table 1: Concentration distribution of main contamination substances of Hengshui Lake area.

Monitoring spots	pH	COD (mg/L)	Ammonia nitrogen (mg/L)	Phoshporus (mg/L)	TN (mg/L)
Yellow River	8.10	3.40	-	0.06	3.02
Main lake	8.04	35.6	0.43	0.013	-
Mini lake	7.90	63.2	15.27	0.032	-
Urban drainage	7.70	69.5	18.77	0.152	-

Hengshui Lake will face increasing eutrophication if nutrition levels in the surrounding environment continue to rise. The degree of eutrophication of the water around Hengshui Lake has already increased. The characteristics of the water in different areas of the lake are shown in Table 1. The level of TN in the Yellow River is about 3mg/L; and in water from upstream and surrounding areas Ammonia nitrogen is over 10mg/L. In order to avoid the water of the lake being contaminated, an area named Jizhou Mini Lake was separated from the main lake to collect this nutrient-loaded water (Figure 3).

After Hengshui Lake was enlarged to create the current reservoir, about 20,000 local residents lost their land, with more to follow in future stages of the water transfer scheme. These people still live in the lake area as they have done for many years because they did not want to go other places. Drainage and waste from their new settlements is freely released into the lake and, as a result, eutrophication has increased. Tourism has also been developing rapidly in the area in recent years, bringing many people to visit Hengshui Lake. There is a risk that this may lead to further increases in the discharge of nutrient-laden waste, further increasing eutrophication in the lake.

Figure 3: Jizhou Mini-Lake is on the right and the main lake is on the left.

Water Scarcity

Hydraulic engineering has enabled the distribution of water in previously arid regions elsewhere in China by trapping flows upstream. This has led to sudden and large scale decreases in water availability in relatively heavily populated areas downstream, such as Hengshui Lake. At the same time, populations have also been growing in this area. Densely populated areas near the lake now cover around 292.7/km^2.

Due to shortages of surface water, supplies for industry, agriculture and domestic use in Hengshui city, have been taken from deep groundwater sources. Groundwater extractions for these purposes have taken place at a rate of 1637 million m^3 annually. The over-extraction of groundwater led to the formation of an underground unhydrous funnel area underneath this city. The whole north plain region of China has undergone a similar over-extraction of groundwater, and resulting funnel effect. However, Hengshui city region is the most extreme case.

Water shortage has now been recognized as one of the biggest threats to sustainability in the Hengshui region. The average consumption of water per capita in this region is currently only 185m^3, this corresponds to 1/13 of the average for China, and 1/26 of global average per capita consumption. In order to halt the serious problem of groundwater depletion and shortage in Hengshui region, the government has already refocused its attention onto the surface water of Hengshui Lake.

It was in recognition of the above problems that the government supported the proposal to establish a national nature reserve at Hengshui Lake in 2003. Although this did not lead directly to a solution to the problems of water pollution and scarcity, it added a new dimension to the functions and management challenges of the lake.

Future Projections of the Water Scenario of Hengshui Lake

The requirements for water from the lake to provide drinking water for the cities and for cooling of the power plant are already mentioned in the section above. On an annual basis, these will be not less than 36 million tons for drinking water and 12 million tons for cooling at the power plant. Other requirements for agricultural irrigation, as already mentioned, are in addition to these amounts.

The projections that are examined in this section of the paper concern the effects of the creation of the reservoir and nature reserve at Hengshui Lake, amid the continuation of the pressures and trends identified in the previous section. The establishment and development of the reservoir of Hengshui Lake is creating some problems, both for the surrounding human population and also for the ecological environment, although it also brings great benefit for this region. The main issues are as follows:

Displacement of the Peasant Population

The dispossessed population is projected to reach 65 thousand people following the enlargement of the reservoir. This loss of land will compel peasants to change their old lifestyle of farming and seek alternative lifestyles. The peasants will continue to live by the lake, and to rely on the wetland to provide them with environmental services.

Loss of Water Resource

Water losses from the reservoir through evaporation and leakage will increase due to the larger surface area, across a still relatively shallow depth. Losses from natural transpiration are predicted to amount to around 52.52 milion tons.

Reduced Quality of Agricultural Land

Hengshui region is a low-lying region where rising ground water-levels and an increased volume of surface water can be projected to result from the enlargement of the reservoir. Because of these changes, the surrounding farmland areas will decrease in area, and pressures on the farming land will increase correspondingly.

Eutrophication and Ecosystem Degeneration

These threats can be projected to remain in Hengshui Lake, due to the continued application of nutrients, and the dense population. Peasants in the conservation area and the increasing number of visitors there will also be an important source of nutrient inputs, increasing current trends in eutrophication and degeneration of water quality.

Integrated Management of Hengshui Lake Reservoir and Wetland Nature Reserve

The primary planning on integrated management of Hengshui Lake has already been done, and some essential governing principles for management have been put forward (Urban Programming Institute of Beijing, 2004). An important goal is that through this improving management approach, the participation of the local peasant population can be included, and their livelihoods and the local economy will be supported within the management of Lake ecosystem restored.

Pollution Management and Control

For the up-stream polluted water from densely populated areas, some other natural and constructed wetlands will need to be established for natural water purification. The establishment of a water treatment plant for combined use with agricultural irrigation works will enable the extraction and use of water containing high levels of nutrients. Treated water and sewage will all go into Jizhou mini-lake, and will then be taken off for irrigation or direct use in farm agriculture. Meanwhile, part of the water will be passed through the surrounding wetlands for purification. The water coming out of the wetland that is of poor quality can then be used for the electric power plant as cooling water. Some of the water can still be allowed into the main body of the lake, within the limits of its natural purification capacity for the removal of pollution.

Ecological Restoration

The ideal conditions for ecological restoration are to improve all functions and uses of the Hengshui Lake wetland. This includes also the development of a harmonious balance between the ecosystem and the population, not only in contamination control, regulation of water levels and ecotourism development, but also in the use of the natural resources. Ecological restoration of the wetland is not focused on efforts to preserve the natural environment only. An ecosystem facing eutrophication could not be protected through the creation of a nature reserve alone. This is because the restoration of natural processes can be expected to increase the levels of nutrition in the ecosystem, and might actually accelerate the effects of eutrophication. That is why the ecosystem should not only be conserved, but should also be protected through sustainable use and transfer of nutrients from the wetland water, in such a way as to restore the balance of processes, and to enable some nutrient material to circulate within the ecosystem.

Transfer of Biomass Products

The development and transfer of biomass products from the wetland ecosystem is an important part of the conservation and use of Hengshui Lake wetland ecosystem. Biomass products that can be obtained from the wetland include cattail and reeds used for weaving, feed for domestic animals, aquaculture, fish and fish food, a variety of algae used for the cultivation of edible snails, and many others. Such products can be supplied to tourists and the local economy. The transfer of biomass from the lake will also decrease nutrient levels and reinforce the capacity of the wetland ecosystem to remove contamination and to remain free from eutrophication.

Ecotourism Development

As a strategy for economic growth in Hengshui Lake to improve the livelihoods of local people, ecotourism should be developed to receive five thousand visitors daily. This strategy will lead to the creation of employment opportunities in fishing, entertainment services and other activities on land.

Establishment of a Mechanism for Integrated Management

The management of the nature reserve aims to maintain, increase and protect biodiversity and water quality, as well as to develop the local economy through creating scenic spots for ecotourism. Management facilities will be built and scientific monitoring studies will be carried out. Conservation managers will be recruited and trained. The introduction of new technologies will be studied through national and international cooperation. Codes and regulations of integrated management will be progressively established and completed.

Public Education and Water-Use Efficiency

The concept of water-use efficiency and the social obligation to save water will be reinforced across the society as a whole. Use of water will be coordinated, and water resources will be protected. Reinforcing public awareness and education in the community will improve the relationship between the population and the conservation area.

Remaining Challenges and Problems for the Management of Hengshui Lake

In practice, it is not easy to realize the objectives described above. There are still a lot of challenges to be faced at Hengshui Lake. The main problem concerns what models of integrated management should be adopted to maximise the restoration of the wetland, while providing environmental services to 65 thousand people within the conservation area and enabling the local community to improve their livelihoods.

Scientific Challenges

The core scientific problem concerns the identification of appropriate means to transfer bio-productivity and remove nutrients from the lake. These questions include how to remove water plants and reduce the use of chemical fertilizers for agriculture generally.

Recently, some simple initiatives have been adopted as follows:

- *Introduction of grass carp into the lake*
  Putting fish fry into Hengshui Lake will lead to the consumption of various plants that grow in large quantities in the lake. The removal of these plants will reduce degeneration from the plants dying and rotting. But many plants will still remain in the lake. These plants can not enter into the food chain, and continue to contribute to eutrophication.

- *Harvesting and processing water plants*
  Protruding water-plants, such as reeds can be directly harvested, processed, used for weaving by local people. But the amount of plants harvested in this way is low because there is not much demand for them. In addition, the local community has been dissuaded from harvesting plants because water diversion from the Yellow River took place during the season when the reeds should be harvested. Because of this, the lake water was too deep for the people to harvest the reeds. This affected their enthusiasm and the volume of the reeds that could be harvested.

- *Reverse succession of the eutrophication ecosystem*
  Breeding larger animals in the shore belt (riparian zones) of the lake should stimulate the wetland ecosystem food chain and lead to the reduction of nutrient level in the lake water.

Management and Policy Challenges

The traditional management of Hengshui Lake through existing government arrangements cannot meet with the rapid development pressures and the burdens created by the many functions of the lake ecosystem. Integrated management models have to be applied in new ways in order to enable the available technologies to be developed, combined and used innovatively. The management mechanism should be able to supply theories and technology to support management activities. It should also be able to respond more rapidly to solve problems. Improvements are needed at every level of the current mechanism. Particularly at the highest level, the process to determine policy and technical support should be enhanced.

The creation of a biosphere reserve at Hengshui Lake can help to provide necessary guidelines for the use of science and technology to protect both the ecosystem and human livelihoods. Hengshui Lake could become a model for new theoretical and applied research on management approaches and technologies. This will promote ecological benefit, social benefit and economic benefit, leading to sustainable development.

References

Guan, Z.J. (2005) "Discussion about releasing from sediment of Miyun reservoir and estimation of inner-source contamination charge", *Beijing Water Resources* 121(2): 28-29.

Mitsch, W.J. and J.G. Gosselink. (1993) *Wetlands*. New York: Van Nostrand Reinhold.

Urban Programing Institute of Beijing (2004) "General planning of Hengshui Lake", Hengshui Lake National Nature Reserve of China: Management of Hebei Province, Qinghua.

Wang, B., Z. Xijing, and Suhualong (2001) "Science search report on Hengshui Lake nature reserve", 12.

Wang, G., X. Cheng, and P. Pu. (2002) "Lake eutrophication control in technology, theory and application", *Journal of Lake Sciences* 14(3): 273-282.

Yang, H. and A.J.B. Zehnder. (2005) "The south-north water transfer project in China - An analysis of water demand uncertainty and environmental objectives in decision making", *Water International* 30(3): 339-349.

Zhuojin, G. (2005) Integrated Management and Conservation of Hengshui Lake Wetland Ecosystem. UNU-INWEH/UNESCO-MAB-IHP International Workshop on Water and Ecosystems: Water Resources Management in Diverse Ecosystems and Providing for Human Needs. 14-15 June 2005. Hamilton, Ontario, Canada.

The Guadiana Estuary

Luis Chícharo, Alexandra Chícharo, Radhouane Ben-Hamadou and Pedro Morais

Faculty Marine Sciences and Environment, University of Algarve, Portugal

Introduction

The Alqueva dam in southern Portugal has been trapping the Guadiana river water since its construction in 2002, causing several changes in the estuary and coastal ecosystems and affecting their use by local populations. The current paper examines the issues raised by changes in flows into the Guadiana estuary ecosystem in light of a global effort to define ecosystem needs for water flow as an integral, life supporting, "ecosystem flow". This essential flow includes the necessary seasonal and inter-annual variation that is needed in varying flow pulses in order to support native species and critically important ecological functions, as well as to provide the sustainable water source that is used by populations.

The current state of knowledge indicates that the impact of dams on ecosystems is profound, complex, varied, multiple and mostly negative (Graf, 1999; Adams, 2000). By storing or diverting water, dams alter the natural distribution and timing of stream flows. This in turn changes sediment and nutrient regimes and alters water temperature and chemistry, with consequent ecological and economic impacts. Reduction in downstream annual flooding affects in particular the natural productivity of floodplains and deltas. Ecosystem alterations result from a significant impact of dams on freshwater biodiversity, which is already under special threat worldwide. Dramatic reductions in bird species are known to occur as a result of dam construction, especially in downstream floodplain and delta areas. Even if some reservoirs provide habitats for birds and other fauna, this often does not outweigh the loss of habitat downstream.

Multiple dams on a river significantly intensify the impact on ecosystems. Sediment entrapment can reach 99% if a cascade of dams is developed. Fish migration is affected even by a single dam, and multiple dams worsen this situation dramatically. In the Northern hemisphere 77% of the largest rivers are affected by dams and on many rivers fully natural reaches are restricted to headwaters. The global impacts of dams for the global water cycle are increasingly recognized.

This review highlights the complexity of the processes that result from the impact of a dam on an ecosystem. It is therefore extremely difficult and rarely possible to predict in precise detail the magnitude and nature of impacts arising from the construction of a dam or a series of dams. The precise impact of any single dam is unique and dependent not only on the dam structure and its operation, but also upon local hydrology, fluvial processes, sediment supplies, geomorphic constraints, climate, and the key attributes of the local biota (Bergkamp et al., 2000). There is therefore no normative or standard approach to address ecosystem impacts and these have to be looked at case by case.

The Guadiana river basin ecosystem suffered along the last five decades the impacts imposed by the construction of more than 100 dams. In 2002, the construction of one of the biggest reservoirs in Europe, the Alqueva dam (250 km^2), was concluded.

The construction of these dams is required in order to respond to an increasing demand for water to support human well-being and to promote regional economic development. However, an ecologically sustainable water management program must always be built upon a foundation of knowledge about the river flows needed to sustain ecosystem health. When the water needs of an estuarine ecosystem are clearly defined by scientists, water managers will be able to find ways of meeting human needs for water while maintaining adequate river flows for the ecosystem.

The Guadiana River Basin and Estuary

The Guadiana river basin, located in the southwestern Iberian Peninsula (Figure 1) originates in Spain at an elevation of 1700m and flows south to the Atlantic Ocean through Portugal. The basin area is 67,500km^2 presenting the 4th largest river basin in the Iberian basin. A typical Mediterranean hydrological regime characterizes the Guadiana river flow. This includes high intra and interannual discharge variation, large floods and severe droughts. This temporal variability is a result of a changeable rainwater supply, averaged around an annual mean of 400-600mm. The current river inflow to the estuary depends on the multiple upstream dam activities. Monthly averaged river flows vary between 20 and 600m^{3}s^{-1}, and can reach 1500m^{3}s^{-1} on very wet years (LNEC, 2001). The river outflow to the Atlantic Ocean passes through the Guadiana estuary, which is situated between the highly touristed regions of the Algarve (Portugal) and sea-side of Andaluzia (Spain). The Guadiana estuary has a maximum width of 550m and its depth ranges between 5 to17m. The variation in residence time -between 5 to 90 days- is mainly dependent on both river flow and tidal activity. Tides are semi-diurnal, ranging from 0.8 to 3.5m; their propagation is limited by falls situated 76km from the mouth at Moinho dos Canais. At the lower estuary there is a nature reserve covering a total of 2089ha at the Saltmarsh of Castro Marim and Vila Real de St António. This natural area gives a high nature conservation value to the region.

Figure 1: Location of the Guadiana river basin (southwestern Iberian Peninsula).

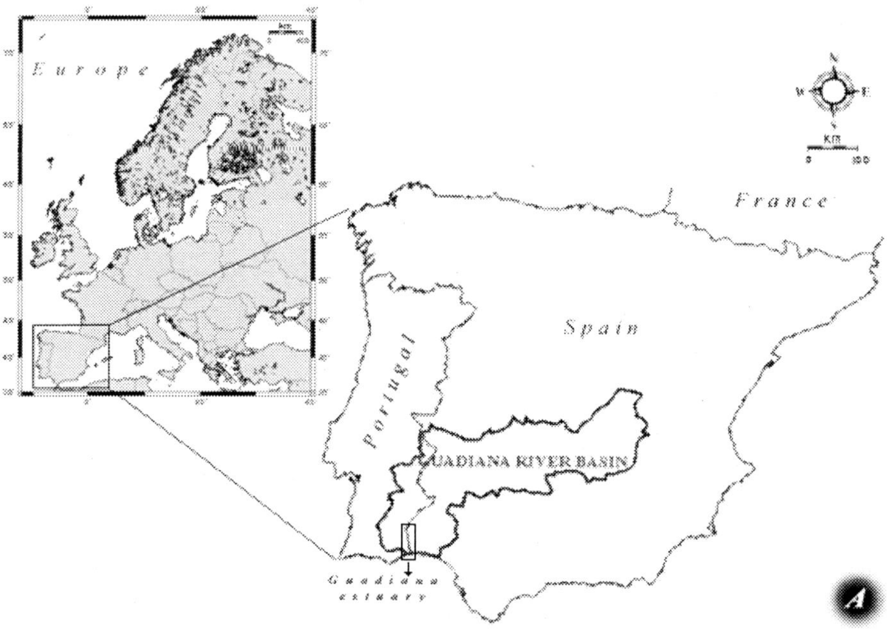

At this lower river basin, water plays a strategic role in the development of several activities such as domestic water supply to half of the Algarve region, fisheries, agriculture, salt making industry and tourism. In addition, water is essential to the functioning of the several ecosystems, as areas in the Guadiana estuary, some being protected areas (Natura-2000 Network, Ramsar Convention, etc.).

Despite data indicating that existing water was enough to ensure human consumption (UNEP, 2004), political pressures led to the construction of the Alqueva dam. These pressures were motivated by increased water demands for agriculture, following changes in the traditional agricultural methods; the need for a strategic water reserve independent from Spain; and the increasing demand for electricity imports and generation in Portugal.

The Alqueva Dam Project

The Alqueva Multipurpose Project was conceived initially in 1957 as part of a regional development project (Alentejo Irrigation Plan). The main objective of this project was the economic development of the semi-arid region around a strategic water reservoir, the Alqueva Dam. The Alentejo region, covering an area of 27,000 km^2, and with a population of around 540,000, is considered one of the most depressed regions of the European Community. The main arguments for the implementation of what is considered the largest artificial lake in Europe were based on the need to combat the growing effects of desertification, to prevent the interannual and monthly fluctuations in rainwater supply and to address social problems in the region. Firstly, the dam was planned to provide water for 110,000ha of intensive irrigation. Coupled to another 9 small dams, an irrigation network with 680km of primary channels, 4400km of secondary channels and 114 pump stations will be created.

On 8 February 2002, the first Alqueva dam gates were closed and the reservoir started to fill the 250km^2 land surface planned to be covered. The dam has a capacity of about 4,150hm^3 with more than 1,000km of perimeter and the capacity to satisfy part of the domestic needs for electricity for the Alentejo region. The project has cost about 450 million euros since 1995, for the dam and the hydroelectrical station. By its conclusion in 2025, it is estimated to cost two billion euros.

The Impacts of the Alqueva Dam

Since their first implementation in the XIXth century, large-dams have brought great benefits to society. However new dam construction has incalculable human costs, as people are displaced and archaeological treasures inundated. Dams also cause the destruction of prime wildlife and biota habitat by submerging the river banks to form the water reservoir. The dam impacts concern not only directly disturbed species but also affect downstream populations. Dams prevent the seasonal flooding that creates species-rich flood plains. Estuaries at the mouths of rivers, deprived of freshwater flow, are often devastated as well.

The Alqueva dam does not constitute an exception, provoking serious consequences for the populations of several threatened species that are protected under European Union law, as well as under several conventions signed by Portugal. Damage to plants are considerable both during and after the dam construction. No less than 1,200,000 trees were cut and at least nine species considered rare, endemic or threatened with extinction (RELAPE) are being severely affected by the filling of the dam. Moreover, terrestrial and bird species are or will be affected around the Guadiana river basin, inducing a real loss for the ecosystem richness in the region

(Chícharo, 2003). In addition to these more local impacts, large scale destruction of habitats for aquatic animals, fisheries and wildlife in the estuary and nearby coastal waters are likely to occur.

Flow Reduction Impacts

The volume of the Guadiana river water stored upstream is increasing since 1950, but after the inauguration of the Alqueva dam, this value increased twofold (Figure 2).

Figure 2: Volume of water stored in the Guadiana Basin from1900 to 2002.

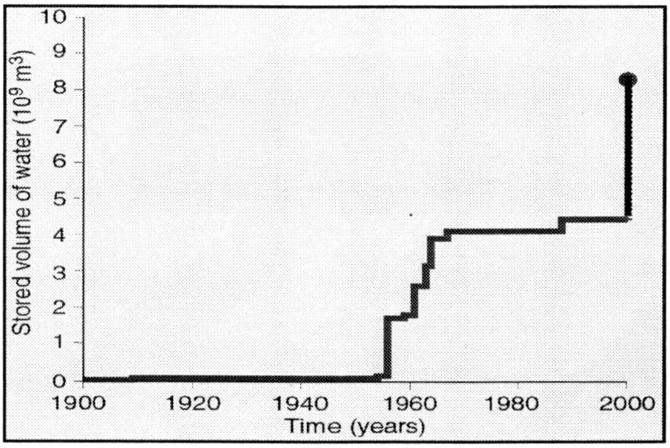

This disturbance in the freshwater outflow caused several impacts in the downstream ecosystems, including modifications in river inflow. The river inflow regime plays a determinant role in estuarine and coastal ecosystems since it is the provider of nutrients and sediments to the estuarine and coastal ecosystems. Also, the reduction of the freshwater flow decreases the ability of the river and estuary to conduct self-purification by exporting excess nutrients to the ocean, which increases the potential risk of eutrophication.

Modifications in the river discharge are responsible for changes in the abundance and distribution of fish species with economic importance and are contributing to the increasing presence of invasive species in the estuary, such as the bivalve clam (Corbicula fluminea) (Chícharo et al., 2001). Coastal fisheries are also being affected by the modification of the freshwater flow, causing a cascade effect on other related economic activities.

Modifications of fish populations and dynamics are mainly a result of the reduction in the freshwater discharge, causing a decrease in the sediments transported downstream. Reduced outwelling from the estuary is visible in the decrease of the coastal area that is influenced by the river plume (Figure 3). These alterations affecting transported sediment alter the grain sizes and deposition of transported sediments and cause loss of adherence to salt-marsh vegetation, which leads to a decrease in salt-marsh areas.

Figure 3: Areas of coastal zone influenced by different Guadiana river plume conditions (dark grey- high sediment content, light grey low sediment content) (adapted from LNEC, 2001).

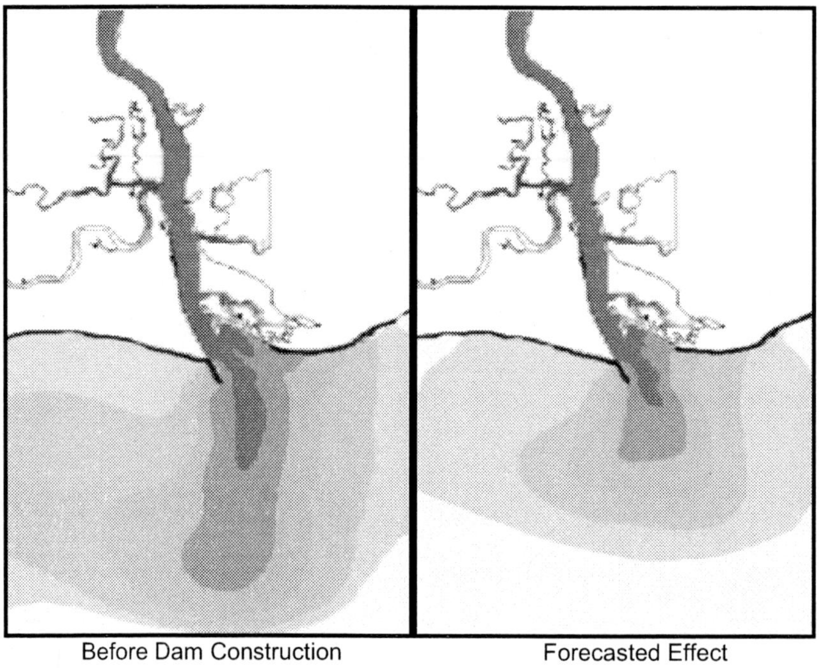

The river plume is used by small pelagic fish species, like the anchovy, as an indication of the proximity of estuaries (Chícharo el al., 2002; Lloret et al., 2004). These species spawn in the coastal areas adjacent to estuaries and the eggs and larvae develop inside the estuary, where better food conditions for larvae and more protection against predators exist. Once the larval and juvenile phase is completed, juveniles swim back to the sea. Therefore, the post-Alqueva decrease in the flow discharge results in a reduced plume, limiting the adult anchovies' chances of reaching the estuary for reproduction. This has consequences for the survival of the eggs and larvae and affects the future size of fisheries; with the consequent social and economic impacts.

The reduction in flow discharge causes a decrease in nutrients in the costal zone and, more importantly, a change in nutrient ratios. The upstream trapping of nutrients affects the N:P:Si ratio and impacts on the ability of diatoms to form their exoskeleton and, therefore, to grow and multiply. The fact that nutrients remain non-consumed by the diatoms increases the possibility of blooms by other phytoplankton groups, some eventually toxic, like the dinoflagellates (Rocha et al., 2002). Moreover, the reduction in the nutrient availability in the coastal zone provokes a decrease in the productivity of the area, affecting all the food web and, more directly, the planktivorous fish species, like the anchovy and sardine (Figure 4). Less available food will support less fishes, which is reflected by a reduction in coastal fisheries catches (Figure 4).

Figure 4: Relationship between Guadiana outflow to the coastal zone and tonnes of *Sardina pilchardus* (planktivorous fish) captured (Chícharo, unpublished data).

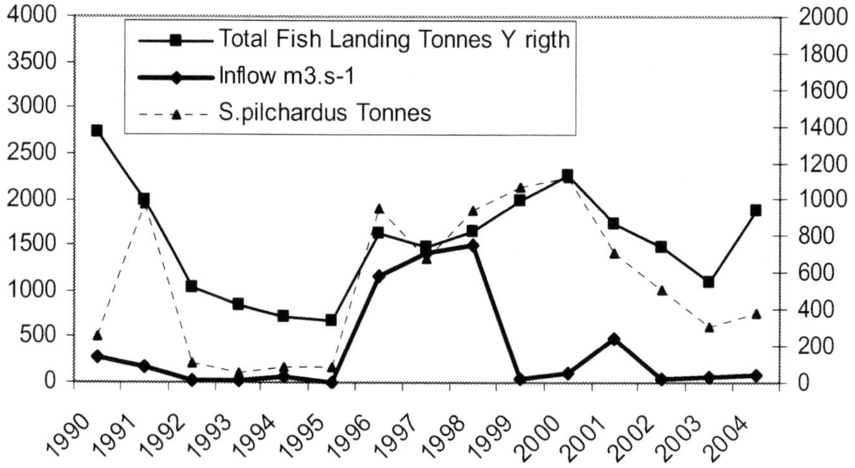

The changes in the coastal fisheries are not just in quantity but also, and more importantly, in terms of the fish assemblages (Figure 5). In fact, there is clear a distinction between years with high river discharge, where planktivorous species dominate, and years with low high inflow, dominated by demersal carnivorous species, indicating a shift from a planktonic to a benthonic food web (Figure 5 and Table 1).

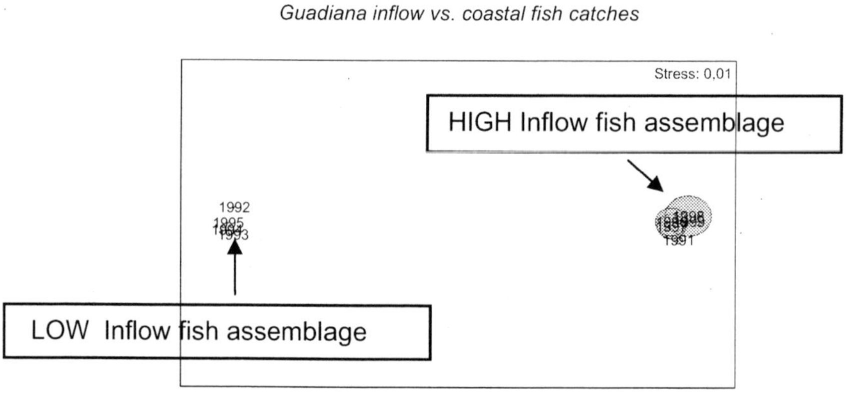

Figure 5: Non metric representation of Multi-Dimensional Scaling (MDS) ordination of log transformed abundances of fish species captured in adjacent Guadiana coastal area between 1990-2000 based on Bray-Curtis similarities, evidencing the existence of clear distinct coastal fish assemblages according to the river inflow to the Guadiana estuary and coastal area.

Table 1: Average abundance and dissimilarity of fish species in Guadiana adjacent coastal area grouped by major groups of years (average high inflow years and low inflow years).

Planktivorous

HIGH INFLOW — LOW INFLOW

Species	Av.Abund **(H)**	Av.Abund **(L)**	Av.Diss	Diss/SD	Cum.%
Sardina pilchardus	885,95	278,82	1,16	2,10	8,78
Engraulis encrasicolus	12,15	9,28	0,84	0,82	15,19
Sparidae n.id	38,40	85,22	0,66	1,58	20,20
Trigla lucerna	6,90	9,08	0,55	2,43	24,37
Diplodus sargus	10,08	23,00	0,52	1,59	28,32
Scomber scombrus	46,13	47,78	0,50	2,16	32,10
Trachurus trachurus	3,45	2,04	0,48	1,34	35,79
Mullus surmuletus	3,88	5,68	0,48	2,46	39,47
Pagellus errythrinus	15,48	26,06	0,47	1,45	43,06
Pagellus acarne	9,10	20,58	0,47	1,56	46,64
Pagrus major	4,35	9,22	0,42	1,84	49,85

Demersal carnivorous

These modifications in the fish assemblage composition are crucial because they affect the fishermen's activities and income, therefore having social and economic impacts on human populations.

Management Tool

The Guadiana estuary and coastal zone is a site where the effects of the construction of dams and their impacts on the ecosystem can be observed and tested. It is clear that the answers to local problems of providing water to populations for their activities require careful planning in order to avoid the transference of ecological problems downstream. Without such care, interventions can lead to the creation of a new set of difficulties for other populations that become limited to their access to water for traditional uses, such as fisheries. When the initial planning to reduce this type of impacts has failed, or where it never existed, mitigation measures are crucial to allow the maintenance of healthy ecosystems.

When looking for the appropriate solution, Ecohydrology and the use of Phytotechnologies appear as the most adequate tools to restore and maintain water quality. These techniques have been used mainly in freshwater ecosystems such as for the Pilica River Basin (central Poland) (UNEP IETC., 2002 & 2003) and their use in estuaries and coastal zones is still in an initial phase. Managing the estuarine ecosystem, due to its complex dynamics, requires a profound knowledge of the system and its functioning.

The data already collected suggested the efficacy of several methods to control the expected increases in eutrophication and toxic algal blooms that result from the decrease in the dilution of fertilizers. The existence of small periods of freshwater discharge seems to have a stimulating effect on the phytoplankton community, multiplying and increasing its diversity, promoting a bottom-up effect on the development of zooplanktonic species. This will then have a top-down effect on the phytoplankton community, reducing the risk of algal blooms (Chícharo, 2004).

Also, phytotechnologies represent important tools in the Guadiana area. The reconstruction of salt-marsh areas will provide a barrier that will improve the nutrient trapping and reduce the eutrophication risk (salt-marshes were reduced because the changes in the grain size of sediment transported downstream).

The effects of hydrology on biota are clearly observed in the effects upon the fish assemblages. Control of the river discharge is crucial not only in terms of the quantity of the water to be released, which needs to be enough to allow the development of a sustainable food web for the fish, but also in terms of the period where such release occurs. If water is released abundantly in the fish species post-spawning period, eggs and larvae may be flushed out of the estuary and areas surrounding the river mouth, with implications for the success of the recruitment and future catches by fishermen. This flow recommendation is also important, not only to fisheries but also to general ecosystem functioning, since it may not be constant in time. In fact, "intensity-duration-date" combinations should be dressed to imitate the functions of either low or high or flood natural pulses.

Based on our observations and experiments an ecohydrological model for the Guadiana estuary was developed that allows users to test different scenarios of impacts and their consequences for the ecosystem (Wolanski et al., 2004), which will be an important management tool for testing future scenarios of water usage in the Guadiana area. This model could be coupled with the IRAS model (Brandão & Rodrigues, 2000) to join the different scenarios encountered by pure hydrological managers, including Spanish water management, together with our ecological modelling of the downstream Guadiana estuary.

The use of Ecohydrological solutions for estuarine and coastal ecosystems is being demonstrated to managers and scientists through the organization of courses and seminars in the Guadiana area, where representatives from the municipalities and management institutions have participated. Moreover, the Guadiana system was considered as a priority area for the implementation of a demonstration site by the Scientific Advisory Committee of the Ecohydrologic Programme of UNESCO. The implementation of this "demo site" will further increase the collaboration with national and regional institutions managing the Guadiana system.

Conclusion

As a conclusion, a decline in available water resources induced by increased demand from a multiple panel of end-users confronts water managers with a dilemma that is often resolved by the implementation of dams. This solution generally responds to the challenge of accumulating water, but also induces a loss of ecosystem performance that has its own direct and indirect negative impacts on human well-being. As a solution for the implementation of future dams, the ecological impacts and solutions of the Guadiana estuary study case should be considered, in order to improve the sustainable use of the ecosystem. For pre-existing dams, an Integrated Water Management regime should be implemented. This can be created through the ecohydrology approach, which focuses on understanding the ecosystem functions and needs. The objective is to develop a new process for implementing ecosystem flows that emphasizes collaboration amongst scientists and water managers, working together to integrate human and ecosystem needs for water. This comprehensive but realistic management approach will guarantee the necessary water supply for both the human population and for the ecosystem, inducing a generalized human well-being.

References

Adams, W. (2000) *Downstream impacts of dams. World Commission on Dams* available from http://www.dams.org/docs/kbase/contrib/soc195.pdf .

Bergkamp, G., M. McCartney, P. Dugan, J. McNeely, and M. Acreman. (2000) *Dams, Ecosystem Functions and Environmental Restoration. World Commission on Dams* available from http://www.dams.org/docs/kbase/thematic/tr21main.pdf.

Brandão, C. and R. Rodrigiues. (2000) "Hydrological Simulation of the International Catchment of Guadiana River", *Physics and Chemistry of the Earth(B)* 25(3): 329-339.

Chícharo, L., M.A. Chícharo, E. Esteves, P. Andrade, and P. Morais. (2002) "Effects of alterations in fresh water supply on the abundance and distribution of Engraulis encrasicolus in the Guadiana Estuary and adjacent coastal areas of south Portugal", *Journal Ecohydrology and Hydrobiology* 1(3): 195-200.

Chícharo, M.A., L. Chícharo, H. Galvão, A. Barbosa, M.H Marques, J. P. Andrade, E. Esteves, C. Miguel, C. Gouveia, and C. Rocha. (2001) "Status of the Guadiana estuary (South Portugal) during 1996-1998: an ecohydrological approach", *Aquatic Ecosystem Health and Management* 4: 73-90.

Chícharo, L. (2003) "Is big beautiful? Alqueva reservoir (Portugal): the largest artificial lake in Europe", *(Societas Internationalis Limnologiae) SIL news* (39): 1-4.

Chícharo, L. (2004) "Estuarine & Coastal Areas: How to Prevent Degradation and Restore (Chap. 13)", in M. Zalewski and I. Wagner-Lotkowska, eds., *Integrated Watershed Management - Ecohydrology & Phytotechnology – Manual*, Japan: UNEP - UNESCO – IHP, pp. 202-208.

Graf, W.L. (1999) "Dam Nation: A Geographic Census of American Dams and Their Large-Scale Hydrologic Impacts", *Water Resources Research* 35(4): 1305-1311.

Lloret, J., I. Palomera, J. Salat, and I. Sole. (2004) "Impact of freshwater input and wind on landings of anchovy (Engraulis encrasicolus) and sardine (Sardina pilchardus) in shelf waters surrounding the Ebre (Ebro) River delta (north-western Mediterranean)", *Fisheries Oceanography* 13(2): 102-110.

LNEC (2001) Estudo das condições ambientais no estuário do rio Guadiana e zonas adjacentes. 2ª fase: Relatório de síntese. Relatório 266/01-NEt. LNEC, pp. 80.

Rocha, C., H. Galvão, and A. Barbosa. (2002) "Role of transient silicon limitation in the development of cyanobacteria blooms in the Guadiana estuary, south-western Iberia", *Marine Ecology Progress Series* 228: 35-45.

UNEP IETC (2002) Guidelines for the Integrated Management of the Watershed - Phytotechnology and Ecohydrology. Freshwater Management Series No. 5.

UNEP IETC (2003) Phytotechnologies. A Technical Approach in Environmental Management. Freshwater Management Series No. 7.

UNEP (2004) *Freshwater in Europe – Facts, figures and maps.* UNEP/DEWA~Europe. Switzerland. pp. 92.

Wolanski, E., L. Chícharo, M. Chícharo and P. Morais. (2004) An estuarine ecohydrology model: The Guadiana estuary case. UNESCO report. pp. 28.

The State of Gujarat

Rajiv K. Gupta

Government of Gujarat, India

Challenges in Water Resources Management and some Success Towards Solutions in the State of Gujarat

"May the waters that are in the sky, or those that flow (on the earth), those (whose channels) have been dug, or those that have sprung up spontaneously, and that seek the ocean, all pure and purifying, may those divine waters protect me here (on earth)."

Wilson, H. H., 1978, Rigveda Samhita, Volume IV, 333

Introduction

This couplet from an ancient Indian scripture succinctly epitomises the eternal importance of water for human life. The last decade of twentieth century has witnessed the process of emphasizing growing shortages of fresh water resources in the world and the need for urgent solutions, both in terms of judicious management of supplies and efficient response to demands. One sixth of Indian territory is drought prone, and one eighth liable to flooding. About 80% of the surface water of the rivers goes to the sea unutilised while the country reels under the flood-drought-flood syndrome. Man-made activities are changing the morphology of regions and river regimes. The aim of this paper is to analyse important planning, institutional, legal and participatory issues confronting integrated water resources development in the country. This paper focuses in particular on the state of Gujarat – a perennially water scarce state in Western India. Some of the successful examples from this state are highlighted.

Figure 1: Rivers South of Tapi River.

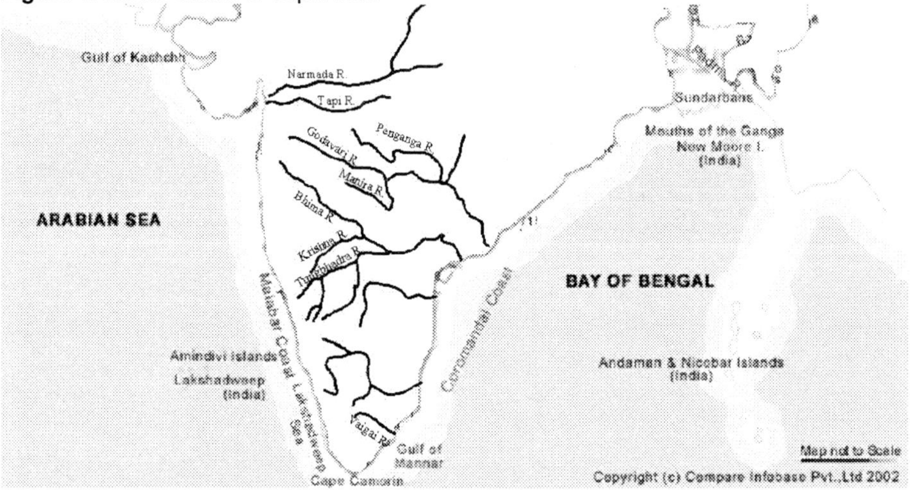

Water Availability and Quality

The total water resource of the country is estimated now at 1953km^3 (GOI, 1999). The water resource of the Ganga-Brahmaputra-Meghna basin is estimated at 1200 km^3, which is 60% of the total water resource flows, while the basin occupies 33% of the geographical area of the country. The water resources in west flowing rivers (Figure 1), south of the Tapi river are estimated at 200km^3, which is 11% of the total water resource, whereas the basin occupies only 3% of the geographical area of the country. The remaining 64% of the area has a water resource of only 553km^3. The estimate of 690.31km^3 has been made by the Central Water Commission as utilisable flow from these figures (CWC, 1998).

Groundwater is an important source of water in many parts of India. The total replenishable groundwater is estimated as 432km^3. Out of this, 396km^3 is estimated as utilisable – 71km^3 (15%) for domestic, industrial and other uses and 325km^3 (90% of the balance) for irrigation. Nearly 50% of irrigation in the country is carried out using groundwater. This water often occurs in the aquifer cones below the zone of water level fluctuation, called static ground water.

> 'In Gujarat, the agricultural, domestic and industrial sectors consume 90%, 9% and 1% water respectively. In 1991, total ground water utilization was 10,416 MCM per year as against total surface water utilization of 9,019 MCM per year. Groundwater is the main source for irrigation contributing 79% of total demand. About 87% of the municipal towns and 78% of villages depend exclusively on groundwater resources. The projected supply – demand scenario for the year 2010 indicates that, in spite of increased surface water utilization up to 23,210 MCM/year, groundwater will continue to be a major source with annual utilization of around 20,940 MCM' (p78) (Vyas, 2002).

Managing groundwater resources has been challenging in Gujarat, a water deficient state of India, where depletion and degradation of groundwater has caused serious concern. Around 30% sub-divisions of the State (for administrative purposes the state is divided into 25 divisions and 225 sub-divisions) are in the 'overexploited' category where the groundwater withdrawal is more than the recharge. Due to overextraction, groundwater tables are falling steadily in Saurashtra (south-western Gujarat), Kachchh (western Gujarat) and North Gujarat. More than 40% of the State's electricity is consumed for extracting groundwater to an average depth of 200 to 300m in some parts. Overdrafting of groundwater or the situation of water mining (as compared to the annual recharge) has caused serious water quality problems like excessive concentrations of fluoride, nitrate and salinity and has thereby led to serious diseases like fluorosis, kidney stones etc. More than 25% villages of the state suffer from water quality problems - North Gujarat being the worst affected, with more than 38 % villages reporting such problems (Table 1). Thus, groundwater tables in state of Gujarat have been drastically going down, creating a challenge for the management of groundwater resources.

Future Water Requirements

Estimates of water requirements have been made for the years 2010, 2025 and 2050 at the national level by many agencies.

Year	Total Water Requirement
2010	694 to 710 km^3
2025	784 to 850 km^3
2050	973 to 1180 km^3

Source (GOI, 1999).

It has been calculated that the total water requirements of the country would be 694 to 710, 784 to 850 and 973 to 1180km^3 by the years 2010, 2025 and 2050, respectively, depending on the low demand and high demand scenarios. Irrigation would continue to have the highest water requirement, between 628-807 km^3 (or about 68% of the total water requirement), followed by domestic water use, including drinking and bovine needs, at about 90-111km^3 (or about 10% of the total water requirement) in the year 2050. The projected water use per capita per year in

Table 1: Water Quality affected habitations in Gujarat.

Sr. No.	District	Number of total habitations	Number of quality-affected habitations				Number of multiple quality problem habitations			
			F	N	S	Total	F+N	F+S	N+S	S+N+S
1	Ahmedabad	723	161	2	101	264	1	16	1	0
2	Amreli	615	59	20	126	205	0	4	0	0
3	Anand	642	91	51	67	209	1	9	6	0
4	Banaskantha	1680	507	69	58	634	9	16	3	4
5	Bharuch	762	24	25	132	181	0	2	4	1
6	Bhavnagar	794	118	38	98	254	2	5	3	2
7	Dahod	1368	281	24	25	330	6	10	9	1
8	Dangs	325	0	0	0	0	0	0	0	0
9	Gandhinagar	363	140	30	7	177	0	0	0	0
10	Jamnagar	765	58	0	232	290	0	4	3	0
11	Junagadh	1032	105	182	157	444	0	3	6	0
12	Kutch	1054	39	1	183	223	1	27	0	0
13	Kheda	1842	450	113	177	740	9	15	10	1
14	Mehsana	649	189	18	107	314	2	12	0	1
15	Narmada	734	35	23	6	64	0	0	2	0
16	Navsari	2062	22	2	59	83	0	0	2	0
17	Panchmahal	1240	407	75	58	540	7	20	11	3
18	Patan	709	327	0	110	437	0	33	2	0
19	Porbandar	184	65	41	40	146	5	4	11	0
20	Rajkot	860	127	230	268	625	1	14	7	0
21	Sabarkantha	2413	513	213	63	789	1	13	10	3
22	Surat	3074	44	69	187	300	2	3	14	0
23	Surendranagar	681	227	23	80	330	3	24	7	5
24	Vadodara	2190	350	84	167	601	4	30	14	1
25	Valsad	3508	2	3	67	72	0	1	0	0
	Total	**30269**	**4341**	**1336**	**2575**	**8252**	**54**	**265**	**125**	**22**

Note: **F**= Fluoride **N**= Nitrates **S**= Salinity/Brackishness/Total Dissolved Solids (TDS)

the year 2050 would be about 725-750m^3, as compared to about 650m^3 at present (GOI, 1999). Therefore, some experts have estimated that by the year 2050, the requirements for water will exceed the available utilisable water resources by around 20% (Varma, 1999). Even presently, availability of water is pretty uneven in both time and space. Precipitation is confined to only about three or four months in a year and varies from 100mm in the Western parts of the state of Rajasthan to over 10000 mm at Cherapunji in state of Meghalaya. Therefore, average water availability at the national level does not ensure that all the basins are capable of meeting their full requirements from their internal resources. It has been estimated that there are six river basins in India where presently per capita water availability is less than 1000 m^3, considered as a scarcity condition by the United Nations. The ethical issue of equity in access to water amongst regions and between sections of the population assumes greater importance in what is foreseen as a delicate and fragile balance between the aggregate availability and aggregate requirement of water.

Integrated Management of Water Resources, Ecosystems, Human Well-Being and Ecosystem Services – Current Progress and Problems

Notable ecosystems in Gujurat include dryland systems in inland areas and extensive mangrove systems along the coastline (Nayak and Bahuguna, 2001). Recent approaches to water management in India have emphasized the use of water basin management approaches. In a basin plan, there is a place for the whole range of structures -large to small. The latter has a particularly important role in rain-fed regions of the country, which will remain starved of irrigation facilities even in the ultimate stage of irrigation development. Large numbers of small local water management projects in rain-fed regions can make a significant improvement in the productivity of such lands and provide succour to millions of poor people, while improving the local environment and regional ecology, thereby leading to sustainable development of those areas.

Large Water Resources Development Projects

Large water resources development projects have frequently been constructed in India in an attempt to address human needs for water. Water resources development was a focal point of India's early Five Year Development Plans with projects like Damodar Valley Corporation, Hirakud, Nagarjunasagar and Tungbhadra etc. to provide irrigation to tide over mounting food deficits and resultant food imports. When the Bhakra Dam was commissioned in early sixties, the first Prime Minister of India, Jawahar Lal Nehru, termed it as "The Temple of Modern India". Of course, it would be difficult to imagine an India without the Bhakra-Pong complex. Its waters provided seven to eight food grains out of ten in every mouthful consumed by all those dependent on the public distribution system in the country till as late as the early nineties (Verghese, 1994). Similarly, the Rajasthan Canal or Indira Gandhi Nahar, as it was renamed in 1984, holds the same importance to the desert state of Rajasthan as the Nile holds for Egypt. It has brought waters of the Bhakra-Pong Dam complex to the desert areas of Rajasthan, irrigating more than a million hectares of land, reversing the trends of out-migration from once parched lands. Recently, even before the formal commissioning of the Sardar Sarovar Project, the waters of the Narmada River, flowing into the main canal through the Irrigation By Pass Tunnel have quenched the thirst of millions of people in traditionally drought-affected regions of Gujarat State.

Experience in the implementation of many large dam projects has shown that their gestation periods were unduly long, leading to cost overruns. The reasons were inadequate planning, insufficient financial allocations in successive development plans, long drawn resettlement and rehabilitation of project affected people. However, other problems have also dogged these projects, they include under-utilization of potential, water logging and salinity due to inadequate drainage and other manmade activities, poor operation and maintenance, lack of canal automation on a large scale etc. Many negative effects on ecosystems from large dams have been reported.

In spite of past problems, large projects are still under consideration to meet water needs in Gujurat and relieve the pressure on groundwater resources. Currently, Gujurat State is looking to the new Kalpasar dam project to fulfill its long-term water and energy needs. The Kalpasar Project, involves the construction of a 660 km long canal system with a maximum discharge of $425m^3$/sec, to convey 5461 MCM of water annually to irrigate 1,054,500ha of land of Southern Saurashtra. It will also provide 900 MCM of water for domestic uses and 500 MCM of water for industrial development in Saurashtra and Kachchh regions. By the creation of a huge fresh water lake and partly a tidal power lake, the Kalpasar will also generate eco-friendly renewable tidal power to the tune of 5880 MW, without the displacement of any communities or the flooding of any land.

Figure 2: Proposed Kalpasar Project.

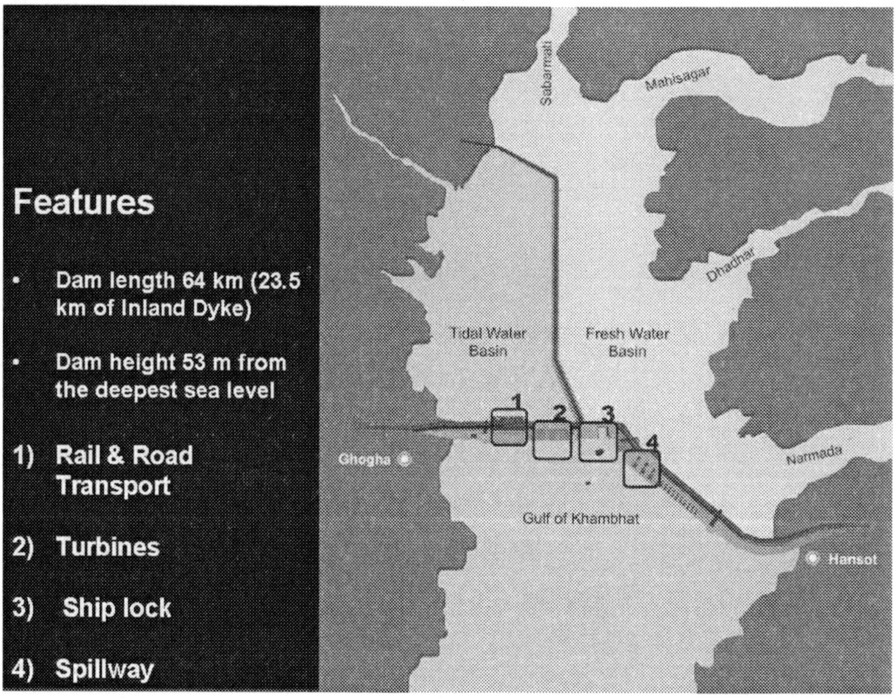

This project is now in the environmental impact assessment phase (see http://www.kalpasar.gujarat.gov.in/). Preliminary studies on the impacts that this project will have on ecosystems consider impacts on mangroves and fisheries (Govt of Gujurat, 1999). The interactions between salt and fresh waters in the extended estuarine system of the Gulf of Khambhat would clearly be affected by the project. The current system of sediment transport, extensive mudbanks and wide range of coastal habitats would also undergo changes. These characteristics make the Gulf of Khambat a particularly important habitat for fish and bird species, such as the near endangered lesser flamingo (Jadhav and Parasharya, 2004).

Smaller Scale Groundwater Management Projects

With the emphasis on large dams and canal systems and also due to socio-economic changes in rural areas (the collapse of traditional power structures, the non-emergence of a cohesive village community in all places and changes in land holdings etc.), the emphasis on tanks and other local resources has waned over the years. The interest and emphasis on local resources also waned because of cultivation, encroachments, deforestation and population pressures in the catchments of the tank systems, their siltation, lack of maintenance and repairs and in several cases, breaches and failures. However, the government of Gujurat has recently returned attention to a combination of both macro and micro-level solutions within the context of watershed development strategies (Gupta, 2004).

The state government has constructed thousands of small-scale check dams, distributed over Saurashtra, Kachchh, Ahmedabad and Sabarkantha region. The objective of the state strategy is to recharge groundwater in depleted aquifers by percolating the stored runoff water which otherwise would have flown downstream. With the people's participation, the State has initiated a successful movement for constructing check dams, known as Sardar Patel Participatory Water Conservation Programme (SPPWCP), wherein a total of 56,000 check dams have been built in last couple of years, providing gross storage of 430 MCM of water. Under this project, sixty percent of the total cost of the work is to be borne by the state government while the remaining cost is shared by the beneficiaries. Non Governmental Organizations (NGOs), voluntary organizations, local leaders etc. are also active in coordinating the activities of the beneficiary groups.

An independent evaluation of the scheme affirmed the efficacy of these small-scale projects as follows (CMA, IIM, 2002):

- Participation of the stakeholders helps them realize the value of water and thereby promoting judicious use and conservation of water.
- Even during a drought year, when the check dams were spilled over on an average 0.92 times, 10257 check dams were able to recharge 138.47 MCM of runoff water which in a normal rainfall year could be 307.71 MCM.
- Hand-pumps which went dry during the previous year were recharged within 7 days of check dam overflowing, and continued to receive water for a period of 3 to 4 months even during scarcity months.
- The number of wells benefited by a check dam ranged from a minimum of 1 well to a maximum of 45 wells, the average being 7 wells.
- 73 percent of the respondents indicated that there was a rainfall after the construction of the check dam which resulted in improving water levels in all the wells and not only those of their own. 60 percent of the respondents reported that their check dams had overflown after construction with an increase in the water levels in their wells of 19 feet on an average.
- In some places, the farmers had sealed their dried wells, but after construction of check dams when their neighboring wells were recharged, they decided to revive their sealed wells. Farmers not having wells on the fields near the newly constructed check dams were planning to construct wells on their fields.
- Even if only crop related benefits are considered, it was possible to recover the investment on a check dam in a very short period of 2 to 3 years.
- Many farmers felt that benefits from the constructed check dams would be substantially higher should there be a normal rainfall.

Local Management of Water Resources Development Projects

One of the challenges faced in the rural water supply and sanitation sector is the sustainability of the systems, which is possible only if the rural community has ownership and is appropriately empowered to shoulder the responsibility and manage them effectively. The Water and Sanitation Management Organization (WASMO) – an independent and autonomous organization in the State of Gujarat has undertaken a demand-driven, decentralized, community ownership and management based water security programme in about 3000 villages (out of 18000) in Gujarat. It is operated by Village Water & Sanitation Committees (VWSC) consisting of villagers. The activities in this programme are focused around several things:

Decentralisation through Pani Samitis (Water Committees)

At the village level, the most important factor is the presence of a strong and functional village institution, the Pani Samiti, a sub-committee of the village panchayat (representative body). The responsibilities of a Pani Samiti include:

- Preparation of a village action plan;
- Timely collection of partial capital cost and full operation and maintenance (O&M) cost, and household water tariff;
- Construction of water supply, sanitation and water harvesting structures;
- Equitable water supply according to designed norms;
- Transparency in fund collection and utilization;
- Ensuring optimal water use;
- Ensuring that increased availability of water will lead to improved hygiene practices;
- Making sure that environmental cleanliness is maintained; and
- Ensuring that the facilities created are well maintained, functional and sustainable.

Capacity Building

Several efforts are made to build up the capacity of those involved in the programme; the village community, ISAs (Implementation Support Agencies), CMSUs (Coordination, Monitoring and Support Units) and the staff of the WASMO head office. Various issue-based workshops are conducted using the services of village-level, state, national as well as international experts. Exposure visits are arranged as well. Emphasis is laid on training of the Pani Samiti members.

Women at the Helm

Women are the providers, users and managers of water in the household. Disseminating information to them about new and improved water supply systems and hygiene practices is thus crucial. The programme ensures that women:

- are Pani Samiti members and signatories of the bank account;
- attend and participate in meetings, workshops and exposure visits;
- see their concerns targeted during awareness programmes and all other activities; and,

- are involved in construction.

Sustainability

In order to instill a sense of ownership and responsibility for the infrastructure created, the village communities contribute in cash, kind or labour following procedures that are decided at Gram Sabhas (village general body meetings). The cost of operation and maintenance are entirely borne by them. Cross-subsidization may be adopted for households unable to pay.

Therefore, the programme has resulted in sustainability of water supply system and improvement in quality of people's life in a large number of villages in the State.

Conclusion

Thus, we have seen that the water demand and supply scenario is very delicately placed in India. With burgeoning population and successive droughts, complicated challenges are likely to be thrown for water management in India. While harnessing the waters of river like Narmada in perennially parched hinterlands of Western India has given some relief and succour to a few millions, it has also led to voices of protests in resettlement and rehabilitation of Project Affected People. While large water development projects continue to be dogged by controversies, innovative efforts of groundwater recharge with people's participation have definitely augmented the capacity to withstand drought years. Government and people's interface in empowerment of rural communities for managing their local water resources and supplies has brought dividends in the State of Gujarat.

References

Central Water Commission (1998) Water and Related Statistics. New Delhi.

Centre for Management in Agriculture (CMA) (2002) Independent Evaluation of Checkdams in Gujarat: Strategies and Impacts, Ahmedabad: Indian Institute of Management.

Government of India, Ministry of Water Resources Development. (1999) Integrated Water Resources Development: A Plan for Action. Report of The National Commission for Integrated Water Resources Development, Vol. 1, New Delhi. (2002) National Water Policy, 2002. New Delhi.

Government of Gujarat. (1999) *KALPASAR Gulf of Khambhat Development Project SPECIFIC STUDIES SUMMARY AND REAPPRAISAL REPORT, Narmada Water Resources and Water Supply Department* available from http://www.kalpasar.gujarat.gov.in/.

Gupta, Rajiv K. (2004) "Water Governance in Gujarat State, India", *International Journal of Water Resources Development* 20(2):131-147.

Jadhav, A. and B.M. Parasharya. (2004) "Counts of flamingos at some sites in Gujarat State, India", *Waterbirds* 27(2): 141-146.

Nayak, S. and A. Bahuguna. (2001) "Application of remote sensing data to monitor mangroves and other coastal vegetation of India", *Indian Journal of Marine Sciences* 30 (4): 195-213.

Varma, C.V.J. (1999) The Rationale of Dam Projects. International Workshop at Antalya, Turkey.

Verghese, B.G. (1994) *Winning The Future: From Bhakra to Narmada, Tehri, Rajasthan Canal*, New Delhi: Konark Publishers Pvt. Ltd.

Vyas, J.N. (2002) "Dams, environment and regional development – harnessing the elixir of life: water for overall development", *Water Science and Technology* 45(8): 71–88.

The Saint John River

Kelly R. Munkittrick[1], R.A. Curry[1], J.M. Culp[2], R.A. Cunjak[1], D.L MacLatchy[1], K.A. Kidd[1], S.E. Dalton[1], D.J. Baird[2] and R. Newbury[1]

[1]Canadian Rivers Institute
[2]National Water Research Institute (Environment Canada)

Introduction

The Saint John River in New Brunswick is typical of many large rivers in southeastern Canada. This paper reviews the major water resource issues in the New Brunswick section of the Saint John basin. Watershed management issues cover a broad range of issues and depend on site-specific aspects of water availability and river quality. The eastern coast of Canada is not typical of the worldwide issue of poor water availability – the basin has an abundant supply of water, the population is relatively small and declining, and there are not major increases in industrial or agricultural needs predicted in the near future. The focus of this paper is on developing a framework for assimilative capacity, focusing on protecting the health of the aquatic ecosystem and ecological services, as well as the availability and quality of water resources for human needs. The objective of the long term studies in this basin is to develop a model for assimilative capacity that integrates human and ecological needs for services.

General Context

The Saint John is a 700km long 7th-order river system that crosses the international border between the US and Canada in the upper third of the basin and forms the border between Canada and the US for >50km. It flows 673km from its headwaters in Northern Maine (USA) through the southeast corner of Quebec (Canada) and down the western portion of New Brunswick into the Bay of Fundy at 45°15' N latitude at the City of Saint John (Figure 1). The Saint John River basin area is the largest in the region at 55,110km^2 (Cunjak and Newbury, 2004), and includes numerous important tributaries such as the Kennebecasis, Nashwaak, and Tobique Rivers in New Brunswick, the Allagash and Aroostook Rivers in Maine, and the Madawaska River in Quebec.

The Saint John River basin lies in the New England/Maritime physiographic province (NE), and drops in elevation a total of 481m from the headwaters in the Appalachian highlands to the estuary (see Cunjak and Newbury, 2004). There are three physiographic regions in the basin (Cunjak and Newbury, 2004 Figure 1): the upper basin includes those waters above Grand Falls, a natural barrier to upstream passage by diadromous fishes. The upper Saint John River basin is sparsely populated and in relatively pristine condition (Cunjak and Newbury, 2004). The central basin refers to that section of the Saint John River between Grand Falls and the Mactaquac Dam (west of Fredericton, and essentially the head of tide). The central basin lies in the western sedimentary plain of the New Brunswick lowlands in a wide smoothly glaciated meltwater valley. This reach is impounded by hydroelectric storage dams for much of its length.

The lower basin refers to the section of the Saint John River between the Mactaquac Dam (west of Fredericton) and Reversing Falls (in the City of Saint John where water flows upstream during rising tides generated from the Bay of Fundy). The total population of the basin is less than 300,000 people, with more than 70% of the population concentrated in the lower basin of the river. The lower basin has little relief and collects water to form three long bays, an inland delta, and several large shallow lakes. Two bedrock sills (Kennebecasis Bay with an 11m drop, and Reversing Falls with a 5m drop) limit the exchange of flows (and saline waters) between the deeper estuarine waters and the Bay of Fundy (Cunjak and Newbury, 2004). The mouth of the river has a tidal height of approximately 7m, and waters in the lower basin become increasingly saline as the river approaches the Bay of Fundy. The brackish water forms a surface layer of varying depth (5-20 meters) over the more saline deeper waters of Long Reach and Kennebecasis Bay (Trites, 1960). The deeper saline layer is much cooler than the surface in spring and summer, with the reverse being true in the fall and winter (Trites, 1960).

Figure 1: The Saint John River basin (adapted from Curry and Munkittrick, 2005).

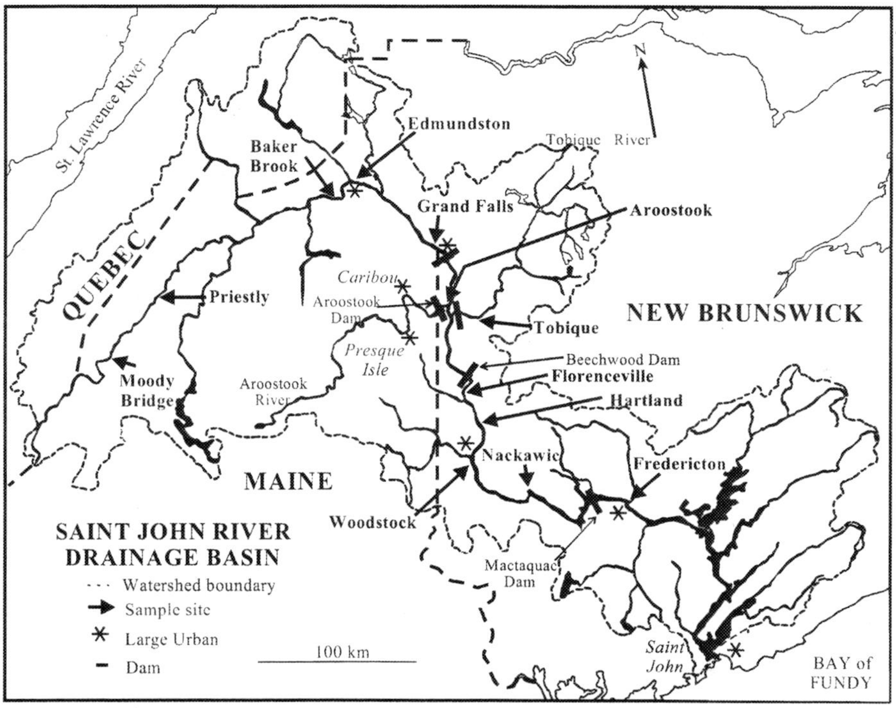

Mean annual discharge for this large river is about 1110m^3/s (from Cunjak and Newbury, 2004). Precipitation through the basin varies with an average annual rainfall of 90cm in the south and 64cm in the headwaters, although a larger portion of the precipitation runs off the headwater region (71%) than in the lower basin (64%) (Cunjak and Newbury, 2004). The precipitation is evenly distributed throughout the year. Much of it is stored in snowpack. The largest runoff (>15 cm/month) occurs when the snow melts in April and May; the highest stages of the river occur in reaches that are prone to ice jams.

Condition and Trends of Water Resources

There are a large variety of types of stressors affecting the river system (Table 1), although water quality is still relatively good because of the size of the river and the relatively small population. Settlements along the river were linked traditionally by river transportation, but the flow of the river is now interrupted by a series of hydroelectric dams; most of these dams have been developed without concerns for fish passage (Table 1). Prior to the construction of the Caribou Dam on the Aroostook River in 1890, virtually all of the 33,300km^2 river system downstream of Grand Falls was accessible to diadromous fishes. Given the paucity of data prior to the building of the first dams, and its large drainage basin, it is possible that the Saint

John River was historically the greatest producer of diadromous fishes in eastern Canada (Cunjak and Newbury, 2004). Today, there are 11 dams in the Saint John River system, including 3 mainstem dams (Mactaquac, Beechwood, and Grand Falls) with valley-wide impoundments in the middle basin above the tidal waters (Figure 1).

Many small communities along the river still rely on forestry, particularly in the upper basin (Maine). The most intensive harvesting for the pulp and paper industry is concentrated in the headwaters of the Tobique, upper Saint John, and Nashwaak Rivers (Cunjak and Newbury, 2004). Wood processing mills are located along the river with the largest plants at Edmundston, Grand Falls, Nackawick, and Saint John (see Cunjak and Newbury, 2004; Curry and Munkittrick, 2005). Potatoes are the most important cash crop in the province with approximately 55,000 acres in production, most of it in the Saint John River valley principally between Grand Falls and Woodstock (see Gray and Munkittrick, 2005). Poultry and hog farms are concentrated in the upper and middle basins. Beef and dairy cattle farming are of similar importance with dairy farms being most common in the lower river valley around Sussex and the Kennebecasis sub-basin. Grain crops such as barley and oats are of moderate importance and largely support the cattle industry. There is commercial farming of blueberries in the highlands and cranberries in the floodplains of the lower valley (Gray and Curry, 2002). Inputs into the river in the area of the City of Saint John include waste from a pulp and paper mill, treated and untreated sewage, a brewery, a paint brush factory, and boat traffic (LeBlanc, 1998). The Saint John harbour receives additional inputs from a large oil refinery, a paper mill and a power plant, as well as treated and untreated sewage (Vallières, 2005).

Water chemistry in this river is very much influenced by the degree and type of land use and the influence of industrial/urban discharges along its corridor. Based on recent sample collections in the main river from the headwaters (Maine) to Fredericton, the following results characterize water chemistry in the Saint John River during the ice-free period (R.A. Curry, CRI, unpubl. data): average pH 7.7 (range 7.3 to 8.1), alkalinity 51.2 mg $CaCO_3$/L (25 to 102), conductivity 154µS/cm (83 to 275), total nitrogen 0.47 mg/L (<0.3 to 1.37), total phosphorous 0.115 mg/L (0.005 to 1.33), dissolved organic carbon 7.1 mg/L (3.2 to 14.0).

Table 1: Types of stressors affecting water quality in the Saint John River studies (adapted from Curry and Munkittrick, 2005).

Stressors	Type	Site	Location (km upstream)	Elevation (m above sea level.)
Forestry, recreation	Main stem	Priestly	625	338
Sewage, poultry processing	Main stem	Baker Brook	447	295
Pulp mill, paper mill, sewage treatment plants, pig production	Main stem	Edmundston	420	136
Potato agriculture, potato processing, hydroelectric dam, no fish passage (natural barrier)	Main stem (reservoir)	Grand Falls	360	135
Forestry, hydroelectric dam (with fish passage)	Tributary (reservoir)	Tobique	325	100
Potato farming and processing, sewage treatment facilities, de-commissioned air force base, hydroelectric dam (no fish passage)	Main stem	Aroostook	329	80
Two food processing plants, potato farming, and sewage treatment facilities, hydroelectric dam (no fish passage)	Main stem	Florenceville	275	55
Presque Isle tributary has starch production industries and potato production fields	Main stem	Hartland	255	50
Large municipal sewage treatment facility, intense potato production, at the head of the flooded reservoir created by the Mactaquac Hydroelectric Dam	Main stem	Woodstock	233	48

Pulp mill, at the head of the lacustrine reach of the reservoir (Mactaquac Lake)	Main stem (reservoir)	Nackawic	185	45
Urban drainage, downstream of the Mactaquac Dam, no fish passage	Main stem	Fredericton	135	9
Agriculture	Tributary	Nashwaak	128	20’
Sewage, artillery training base	Main stem	Oromocto/ Gagetown	107	20’
Forestry, shoreline development, aging or absent septic systems	Tributary	Canaan/ Washedemoak	63	15’ (20’ in lake)
Sewage	Main stem	Grand Bay/ Westfield	14	10’
Agriculture, peat processing, sewage	Tributary	Kennebecasis	41 (Hampton)	20’ at Hampton
Pulp mill, paper mill, sewage, oil refinery	Mouth	Saint John	0	0’

Historically, the Saint John River has been considered one of the more disrupted river systems in Canada, and has been listed as one of the top receivers of contaminants. To date, much of our knowledge about the health of the Saint John River ecosystem comes from assessments of the dramatic decline in populations of Atlantic salmon (*Salmo salar*) that have occurred over the past 50 years (Cunjak and Newbury, 2004). Their declining numbers, from tens of thousands to a few thousand in recent years, are strongly correlated with the construction of dams, particularly the farthest downstream dam which allows no free passage of fishes (Dominy, 1973). A major study of the biology and socioeconomics of the river occurred in the late 1960s and early 1970s (Meth, 1972). Hydroelectric facilities continue to impose significant obstacles to upstream passage and no opportunity for safe downstream passage. High summer water temperatures, especially in the impoundments, stress the fish and restrict growth for many species. Altered flow regimes may affect in-river movements (e.g., migrant smolts, Carr, 2001), and the abundance of piscivorous species, many of them non-native (e.g., smallmouth bass, chain pickerel, muskellunge), provides additional constraints to juvenile salmon production in the river. Finally, there is a potential for 'genetic pollution' by farmed salmon escapees from nearby aquaculture operations in the Bay of Fundy and the Gulf of Maine (DFO, 1999).

Organic loading from food and wood processing in the upper and middle basins of the river exceeds the capacity of the river to assimilate these nutrients (Cunjak and Newbury, 2004). Watt (1973) found that the heavy organic carbon loading from industrial effluents created a zone of extreme heterotrophy that extended from Edmundston to Beechwood in the summer. This condition was exacerbated in the slow-flowing, depositional impoundments where less dissolved oxygen was available to meet the biological oxygen demand (BOD) resulting in low dissolved oxygen (DO) levels. In winter, with low temperatures and reduced aeration under ice cover, the heterotrophic condition extended as far downriver as the Mactaquac Dam. However, in the lower river Watt (1973) concluded that there was no immediate threat of eutrophication to the estuary based on the phosphate measurements and estimates of primary productivity that were calculated in 1972. The situation in the basin has improved markedly since the 1970s, as large industries along the basin have improved their waste treatment facilities. However, contamination continues to affect the aquatic biota of the Saint John River some 30 years after those initial surveys (Figure 2), and there is still a need for installation of sewage treatment facilities for a large portion of the City of Saint John. A variety of intensive studies have been completed or are currently underway in the upper basin (Culp et al., 2003; Galloway et al., 2003; Flanagan, 2003; Gray, 2005; Curry and Munkittrick, 2005), the middle basin (Doherty et al., 2004; Luiker et al., 2004; Luiker et al., 2005), and at the river mouth (Dubé and MacLatchy, 2000; Vallis, 2003; Vallières, 2005). In the estuary, improvements in effluent quality from the pulp and paper mill have been documented (Dubé and MacLatchy, 2000; Dubé and MacLatchy, 2001); however, difficulties in carrying out assessments of the confounded estuarine environment (Vallis, 2003; Vallières et al., 2005) and the lack of baseline studies, make it difficult to assess the present health of the estuary compared to its historical state.

Figure 2: Effluent sources for the New Brunswick portion of the Saint John River basin (from Luiker et al., 2004).

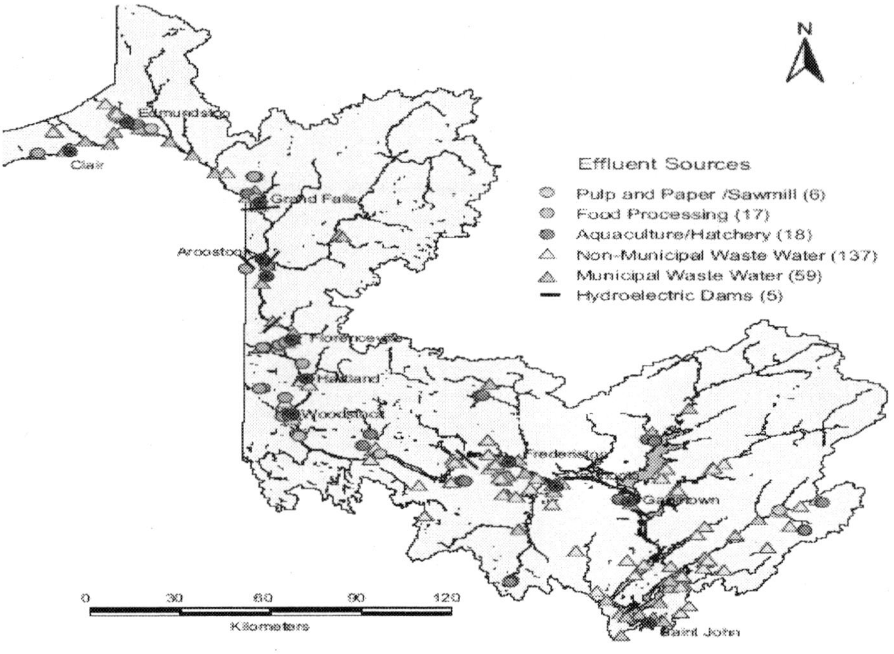

There are concerns about the continuing impacts of nutrients from industry and municipal outfalls, and the impacts of agriculture and food processing plants in the upper basin, although the situation is improving. There are also concerns about the status of sturgeon stocks in the river and the impact of introduced nonnative fish species whose distribution continues to expand (Curry and Munkittrick, 2005), as noted by the recent smallmouth bass and rainbow trout records above Grand Falls (Curry, unpubl. data). There is also concern about possible invasions by zebra mussels from the St. Lawrence River system (Cunjak and Newbury, 2004), and the continued erosion of riparian zones in areas where residences are increasing.

Future Scenarios

It is not anticipated that there will be major changes to the Saint John River system in the near future. The main drivers of ecosystem change are population, development patterns, increased demand, and environmental change (Zalewski, 2000). The population of New Brunswick is slightly decreasing, with a decline of 1.2% between 1996 and 2001 and there are no large industrial developments anticipated in the upper or middle portions of the basin in the near future (http://geodepot.statcan.ca/Diss/Highlights/Page5/Page5_e.cfm). There is a trend towards increasing housing development in the lower basin, with consequent destruction of riparian zone and associated habitat degradation (Jenkins, 2003). There are proposed improvements in sewage treatment facilities, especially for the lower basin; almost 50% of the population of Saint John does not currently have sewage treatment facilities, but these are anticipated before 2015 through funding by municipal, provincial and federal governments (Anon, 2004). Throughout the basin in rural areas there is reliance on septic fields, and sewage and nutrient inputs from cottages and camps are a persistent issue.

The focus of existing studies is on identifying areas of the system that are impacted by human settlement, and on continuing to improve water quality in the system. Major increases in demands for water are not anticipated, and climate change is expected to have only a minimum impact on precipitation for the Saint John River basin (Bootsma et al., 2005). There are ongoing evaluations about proposed changes in forestry operations with a shift towards more plantation forests, there are increasing concerns about groundwater contamination, and there is a heavy dependence of the region on surface water for drinking water sources. There is increasing concern of the impacts of nutrients on the system, and that is the focus of several recent studies (Luiker et al., 2004, 2005), and there are ongoing issues in agricultural and forestry areas with the amounts of sediment being transported into the river systems.

Policy Options

There are a variety of Federal and Provincial regulations that address water quality. At the Federal level, these include the Fisheries Act (1867), the Canadian Environmental Protection Act (1987), The Canadian Environmental Assessment Act (1995) and the Species at Risk Act (2003). Provincially, most water users require permits under the NB Water Quality Act, and there is a recent NB Water Classification Act. The Saint John River in the upper basin is also subject to requirements under the International Joint commission because it forms the international border, and Canada has a number of international agreements and treaties (i.e. Ramsar, Convention on Biodiversity). In addition to the regulations affecting water, there are currently more than 20 community and watershed groups active in the basin and affecting water management decisions (Washademoak Environmentalists, 2003; Dalton and Walsh., 2004).
In Canada, aquatic resource regulations are usually defined by the Fisheries Act and the need to protect fish, fish habitat and fisheries resources. Benthic invertebrate communities are commonly used to define the habitat quality in monitoring programs (Environment Canada, 1998). The levels of contaminants in edible portions of fisheries products are used to assess the protection of fisheries resources. There are well developed programs to assess the contamination of natural resources used for human consumption, and the framework discussed here concentrates on defining the health of the aquatic ecosystem.

While Canada is rich in water resources and the water quality situation is improving in many sectors, there is still an urgent need to develop a framework for integrated water resources assessment. There is a need for a fundamental change in water resources management away from technical and sectoral fixes towards integrated approaches in which the social dimension is central (UN/WWAP, 2003). The assessment of the resource is of fundamental importance and provides the basis for rational decision-making (UN/WWAP, 2003). This need was the basis for the development of a series of watershed studies under the Watersheds and Ecosystems theme of the Canadian Water Network (www.cwn-rce.ca).

Many of the industries within the Saint John River basin are required to conduct monitoring to meet a number of provincial or federal requirements. Historically, the absence of a common philosophy and standardized study design, and the utilization of multiple sampling protocols provides significant challenges to integrating data from multiple monitoring programs. A number of government departments have joined efforts to centralize their data storage on water resources within the province through the NB Aquatic Data Warehouse which will centralize much of the available information on water resources (http://nbwaters.unb.ca). The New Brunswick Aquatic Data Warehouse (NBADW) is a repository of fisheries and aquatic information for the province of New Brunswick. Its purpose is to support ecosystem-based resource management by providing a framework to coordinate the management and sharing of aquatic and fisheries information among provincial and federal governments, NGO's, industry and the public." In addition to data management and distribution, the ADW provides GIS technical support and access to a comprehensive store of data originating from various groups and agencies.

Within the Saint John River system, there are significant gaps and needs for development of institutional, managerial, human, and technological capacity. One of the basic challenges in management is the fractured responsibility, divided between multiple agencies in the federal and provincial government. Within this basin, we have attempted to deal with the major issues of assessment and building a knowledge base by forming an open collaborative network aimed at understanding the ecology of river systems. The Canadian Rivers Institute (CRI) is based at the University of New Brunswick and has a mission to carry out multi-disciplinary basic and applied research focusing on river ecosystems, and their land-water linkages, for the purposes of conservation and habitat restoration. The CRI focuses many of its collaborations on the Saint John River, and a major initiative is the development of a framework for understanding the assimilative capacity of the Saint John River system. The research effort on this watershed has grown into a multi-institutional partnership including federal and provincial government agencies, private industry and NGOs (Figure 3), in a non-competitive, inclusive model that operates without core funding. The growth of this partnership has been gradual, and continues to expand. This multi-institutional partnership functions to provide a direct conduit for recent science into the offices of the policy makers, but it does require a considerable investment of time and energy to develop the trust and connections necessary for the transfer of information to take place. The accumulation of expertise and joint working relationships over a period of several years has been required, but the pace of growth increases with time. This process of slowly influencing policy approaches through science can be successful.

The current studies on the Saint John River are focused on developing a database to facilitate science-based decision-making on water management. The approaches are compatible with the conceptual frameworks offered by the World Water Development Report (UN/WWAP, 2003) that are based on the use of water for basic needs and well-being. Zalewski (2000, 2002) emphasizes using ecosystem properties to understand the system; and the Saint John River studies conducted by the CRI use an effects-based approach (Munkittrick, 2000) to identify and prioritize the areas where there are significant impacts of development and water use on aquatic biota (Galloway et al., 2003; Luiker et al., 2004; Luiker et al., 2005; Curry and Munkittrick, 2005; Gray and Munkittrick, 2005). Follow-up studies in areas of concern are focused on identifying the causative stressors. The overall goal of the studies is to develop a model to understand the current assimilative capacity of the system in order to predict its responses to any additional development.

Figure 3: An outline of the structure of the Canadian Rivers Institute, based at the University of New Brunswick.

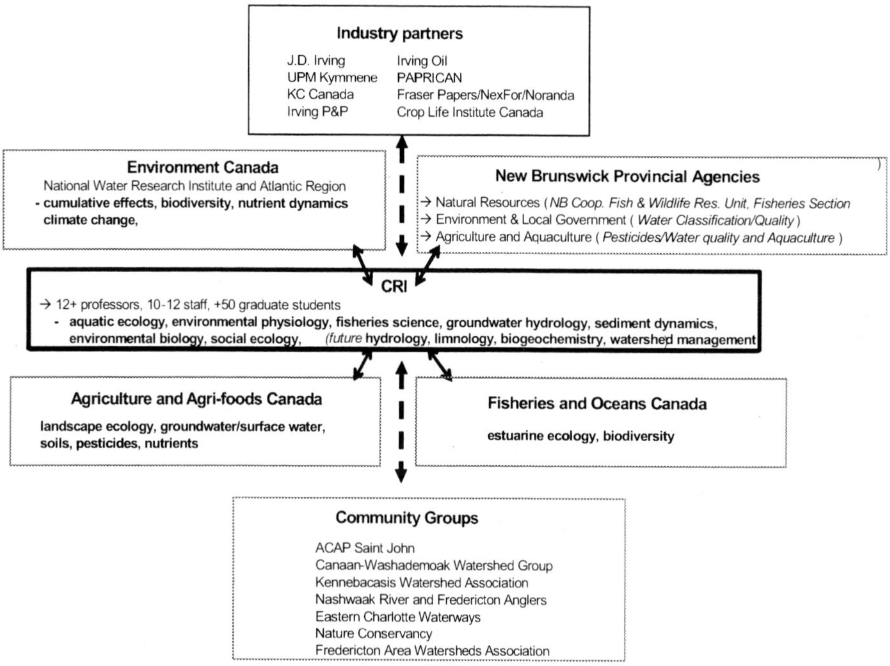

The assessment procedure is focused on a framework for adaptive management that is being developed to understand and protect the sustainability of the river system. Our approach involves the participation of a variety of stakeholders, including provincial and federal governments, industry, public stakeholders and academics. Community stakeholders are key contributors to the process, but the development of significant community capacity for participation requires the education of all stakeholders with relevant policy information, an understanding of terminology and jargon, and the development of a commitment to science-based progress as opposed to the traditional emotional responses of many community groups. Addressing the issue in a multi-stakeholder environment is challenging because each participant brings different perspectives, objectives, study designs and thresholds for interpretation of significance (Munkittrick, 2004). To make progress in this multi-stakeholder environment, we have adjusted our assessment framework through a number of key assumptions that are designed to focus discussions. The key assumptions are that it is necessary to:

a) separate considerations about protecting human health from the issues of protecting ecosystem health and the sustainability of the aquatic ecosystem,
b) separate the desire to detect differences and impacts from interpreting the biological or political significance of those changes,
c) separate decisions about the sustainability of an aquatic system from the acceptability of the changes that are present,
d) accept that impacts and sustainability can coexist,
e) accept that monitoring, adaptive management, and sustainability have temporal constraints and need to be ongoing processes and evaluations.

The effects-based assessment procedure has five steps. The first two steps are to evaluate whether differences in biota exist, and whether these differences are getting better or worse. If the biota are impacted, it is necessary to independently evaluate the sustainability and acceptability of those impacts. Our definition of sustainability is based on the concept that future generations need to inherit an environment at least as good as the present one (Brundtland, 1987). While there is controversy about the criteria used to define sustainability, it may be simpler to define situations that are clearly not sustainable, such as those where key species have been lost or are in danger of being lost. Deciding that a situation is not sustainable means that corrective action is required and needs to be undertaken prior to further evaluation; deciding that it is potentially sustainable means that you can move on to assessing the acceptability of the environmental situation (step 4).

The issue of acceptability of the changes in resident aquatic biota is critical. Like sustainability, the criteria are controversial, but it may be easier to decide what is acceptable rather than what is unacceptable. Unacceptable situations are ones where the possibility of sustainability is threatened, and can occur when key species have been damaged. It is also unacceptable for damage to be increasing in a system, and that would occur if key resources are showing significant impacts that are getting worse over time. Warning criteria are being developed under the Canadian Environmental Effects Monitoring (EEM) program (Kilgour, 2005: see below); to define situations where concerns are sufficient that additional information is required. Reaches of rivers that have impacts that are above warning signs and are getting worse over time means degradation is still occurring. An unacceptable situation is one where species are in danger, changes exist above warning levels and are getting worse over time, or the public stakeholders declare the situation as unacceptable.

Environmental monitoring and assessment can take place at a variety of levels, but there are compromises to be considered during the development of a monitoring program that are related to the sensitivity, reversibility and ecological relevance of the responses, as well as to the time lag to detect them. While ecosystem-level measurements are integrative and ecologically relevant, the time lag to detect them is long, the ability to trace the cause uncertain, and the reversibility challenging. It is critical to define the target that will ensure the protection of the environment. The effects-based assessment approach used in the Saint John River studies involves an assessment of the fish communities, sentinel fish species population characteristics, and measures of benthic community structure and primary productivity. These types of measurements have formed the basis for the development of EEM requirements in Canada. Mandatory EEM programs were developed under the *Fisheries Act* for pulp and paper mills (Walker et al., 2002) and metal mines (Ribey et al., 2002), and are under discussion for a number of other waste dischargers (Kilgour, 2005). The EEM program is a cyclical monitoring program designed to evaluate whether there are receiving environment effects when facilities are in compliance with existing discharge requirements. Their development during the 1990s required the adoption of national standardized study designs and sampling methodologies. The programs were developed through the activity of a wide variety of steering committees, task forces and expert working groups that necessitated extensive philosophical discussions between multiple stakeholders. This approach is consistent with recent trends in regulatory development towards increasing stakeholder involvement and with an increased focus on user-pays.

The standardized approaches and methodologies of the EEM programs are described in guidance documents (www.ec.gc.ca/eem). The approach is based on a framework for cumulative effects assessment (Munkittrick, 2000) that forms the basis for our current focus in the Saint John River basin on using the ecosystem properties and biota to determine the areas where existing impacts are a concern, to prioritize these areas, identify the causative stressors, and to define the assimilative capacity of the system for additional development. The specific focus of the Saint John River studies is on developing an integrated assessment tool for prioritizing areas of concern and for focusing remediation efforts. The philosophical approach will be transferable to many other systems, but it is primarily focused at this time on developed countries.

Zalewski (2000) recommends that an assessment procedure have a sound strategy, flexible tactics and efficient operation. It is not prudent or cost-effective to try to obtain all of the information needed to assess the status of fish and fish habitat in a single sampling event. A cyclical monitoring program is essential for allowing the questions of the relevance of changes to be revisited on a regular basis, and to assess the responses of the system to any environmental change, whether it is anthropogenic or natural. Additional ongoing monitoring is required to ensure that environmental conditions (drought, global warming, large scale eutrophication, etc.) that also impact ecosystems do not change the potential for sustainability. The final (5th) stage of the assessment process is to evaluate the consequences of changes for further development, or in the presence of changing climate.

Conclusion

Integrated watershed management in developed countries is challenged by fractured political responsibility and shifting demographics and land use, and hindered by the absence of a comprehensive ecosystem-based framework for evaluating sustainability. Ecological integrity and sustainability are inherently valuable, but existing watershed efforts usually focus primarily on protecting human health and the human use of ecological resources. Integrating ecological services into watershed management requires an understanding of the accumulated environmental state, the assimilative capacity of the receiving systems, and the potential consequences of future developmental scenarios and changes in climate. The evaluation framework described herein represents an effects-based approach to monitoring, assessment and priority-setting that is aimed at understanding the factors limiting ecosystem services. While the CRI and Saint John River basin approach may offer a useful model, the development of key partnerships and trust takes time and the developmental pace of working relationships needs to be increased to meet the urgent needs in developing countries.

References

Anon (2004) "Water and Wastewater: Saint John in the 21st Century", City of Saint John report. March, 2004.

Bootsma, A., S. Gameda, and D.W. McKenney. (2005) "Impact of potential climate change on selected agricultural indices in Atlantic Canada", *Canadian Journal of Soil Science* 85: 329-343.

Brundtland, G. H. (1987) *Our Common Future*, New York: Oxford University Press.

Carr, J. (2001) "Downstream movements of juvenile Atlantic salmon (Salmo salar) in the dam-impacted St. John River drainage", *Manuscript Report of Fisheries and Aquatic Sciences*, 2573: pp.76.

Culp, J.M., K.J. Cash, N.E. Glozier and R.B. Brua (2003) "Effects of pulp mill effluent on benthic assemblages in mesocosms along the St. John River, Canada", *Environmental Toxicology and Chemistry* 22(12): 2916-2925.

Cunjak, R.A. and R.W. Newbury. (2004) "Atlantic rivers - Canada", in A. C. Benke and C. E. Cushing, eds., *Rivers of North America,* San Diego, C.A.: Acad. Press, pp. 938-980.

Curry, R.A. and K.R. Munkittrick. (2005) "Fish community responses to multiple stressors along the Saint John River, New Brunswick, Canada", in J. N. Rinne, R. Calamusso and R. Hughes, eds., *Changes in large river fish assemblages in the North America: Implications for management and sustainability of native species*: North American Journal of Fisheries Management.

Dalton, S.E. and J.M. Walsh. (2004) "NB Watershed Groups Data Management Survey Report", Unpublished report to the Sciences and Reporting Branch, NB Department of Environment and Local Government Canaan-Washademoak and Fredericton Area Watersheds Associations.

DFO (1999) "Interaction between wild and farmed Atlantic salmon in the Maritime Provinces", DFO Maritimes Regional Habitat Status Report 99/1E, Dartmouth, Nova Scotia.

Doherty, C.A., R.A. Curry and K.R. Munkittrick. (2004) "Adult white sucker show limited mobility near point source discharges in a large Canadian river", in D.L. Borton, T.J. Hall, R.P. Fisher and J.F. Thomas, eds., *Pulp & Paper Mill Effluent Environmental Fate & Effects,* Lancaster, PA.: DesTech Publications, pp. 123-132.

Dominy, C.L. (1973) "Recent changes in Atlantic salmon (*Salmo salar*) runs in the light of environmental changes in the Saint John River, New Brunswick, Canada", *Biological Conservation* 5: 105-113.

Dubé, M.G. and D.L. MacLatchy. (2000) "Endocrine responses of Fundulus heteroclitus to effluent from a bleached-kraft pulp mill before and after installation of reverse osmosis treatment of a waste stream", *Environmental Toxicology and Chemistry* 19(11): 2788-2796.

Dubé, M.G. and D.L. MacLatchy (2001) "Identification and treatment of a waste stream at a bleached kraft pulp mill that depresses a sex steroid in the mummichog (Fundulus heteroclitus)", *Environmental Toxicology and Chemistry* 20(5): 985-995.

Environment Canada (1998) Pulp and paper technical guidance for environmental effects monitoring. Environment Canada Report EEM/1998/1, Ottawa ON.

Flanagan, J.J. (2003) "The impacts of fine sediments and variable flow regimes on the habitat and survival of Atlantic salmon (Salmo salar) eggs in some New Brunswick streams", M.Sc. thesis, University of New Brunswick, Fredericton, N.B.

Galloway, B.J., K.R. Munkittrick, S. Currie, M.A. Gray, R.A. Curry and C.S. Wood. (2003) "Examination of the responses of slimy sculpin (Cottus cognatus) and white sucker (Catostomus commersoni) collected on the Saint John River (Canada) downstream of pulp mill, paper mill, and sewage discharges", *Environmental Toxicology and Chemistry* 22(12): 2898-2907.

Gray, M.A. and R.A. Curry. (2002) "Stream Fish Responses to Blueberry Production: searching for indicators of human-induced stress", Report to the Eastern Charlotte Waterways, Inc. New Brunswick Cooperative Fish and Wildlife Research Unit, Fisheries Report No. 02-06.

Gray, M.A. and K.R. Munkittrick. (2005) "An effects-based assessment of slimy sculpin (Cottus cognatus) populations in agricultural regions of northwestern New Brunswick", *Water Quality Research Journal of Canada* 40(1): 16-27.

Gray, M.A., R.A. Curry and K.R. Munkittrick (2005) "Impacts of nonpoint inputs from potato farming on populations of slimy sculpin (Cottus cognatus)", *Environmental Toxicology and Chemistry* 24: 2291-2298.

Jenkins, R. (2003) "Evaluation of the Canaan-Washademoak ecosystem: toward conservation and sustainable management through water quality monitoring and riparian zone assessment", Cambridge-Narrows, NB.: Washademoak Environmentalists.

Kilgour, B.W., K.R. Munkittrick, C.B. Portt, K. Hedley, J. Culp, S. Dixit and G. Pastershank (2005) "Biological criteria for municipal wastewater effluent monitoring programs", *Water Quality Research Journal of Canada* 40(3): 374-387.

LeBlanc, K. (1998) "Effects on reproductive endocrine function of mummichog (Fundulus heteroclitus) and goldfish (Carrasius auratus) exposed to Saint John, New Brunswick harbour waters", M.Sc. thesis, University of New Brunswick, Saint John, N.B.

Luiker, E., J.M. Culp, K. Heard, and R.A. Curry. (2004) "Determination of the Nutrient Status and Effect of Nutrient Loading on the Saint John River", New Brunswick Cooperative Fish and Wildlife Research Unit Report No. 04-03 NB Environmental Trust Fund.

Luiker, E., J.M. Culp, L. Noel, E. Foster, D. Hryn, and R.A. Curry. (2005) "Determination of the Nutrient Status and Effect of Nutrient Loading on the Saint John River - Year 2", New Brunswick Cooperative Fish and Wildlife Research Unit Report No. 05-04 NB Environmental Trust Fund.

Meth, F.F. (1972) "Ecology of the St. John River basin, V. Status of the estuary fisheries", A report prepared for the St. John River Basin Board, May, 1972.

Munkittrick, K.R. (2004) "The evolution of study approaches with pulp mill effluents, 1991-2003", in D.L. Borton, T.J. Hall, R.P. Fisher and J.F. Thomas, eds., *Pulp & Paper Mill Effluent Environmental Fate & Effects.*, Lancaster, PA: DesTech Publications, pp. xv-xxvii.

Munkittrick, K.R., M. McMaster, G. Van Der Kraak, C. Portt, W. Gibbons, A. Farwell and M. Gray. (2000) *Development of Methods for Effects-Based Cumulative Effects Assessment Using Fish Populations: Moose River Project.*, Pensacola, FL.: SETAC Press.

Ribey, S.C., K.R. Munkittrick, M.E. McMaster, S. Courtenay, C. Langlois, S. Munger, A. Rosaasen, and G. Whitley. (2002) "Development of a monitoring design for examining effects in wild fish associated with discharges from metal mines", *Water Quality Research Journal of Canada* 37: 229-249.

Trites, R.W. (1960) "An oceanographical and biological reconnaissance of Kennebecasis Bay and the Saint John River Estuary", *Journal of the Fisheries Research Board of Canada* 17(3): 377-408.

UNESCO-WWAP. (2003) Water for people, water for life: World water development report, Barcelona: UNESCO and Berghahn Books.

Vallières, G. (2005) An effects-based assessment of the health of fish in a small estuarine stream receiving effluent from an oil refinery. MSc Thesis, University of New Brunswick, Saint John. 172 p.

Vallières, G., K.R. Munkittrick, and D.L. MacLatchy (2005) "Assessing fish community structure in relation to water quality in a small estuarine stream receiving oil refinery effluent", *Marine Pollution Bulletin.*

Vallis, L. (2003) The potential use of rock gunnel (Pholis gunnellus) as a bioindicator in Saint John Harbour, New Brunswick, Canada. MSc Thesis, University of New Brunswick, Saint John . 97 p.

Vallis, L., D.L. MacLatchy and K.R. Munkittrick. (2005) "The biology of rock gunnel (Pholis gunnelus) along the Atlantic coast of Canada, and their potential for use as sentinel species for monitoring environmental quality. Submitted to *Journal of Fish Biology*".

Walker, S.L., K. Hedley and E. Porter. (2002) "Pulp and paper environmental effects monitoring in Canada: An overview", *Water Quality Research Journal of Canada* 37(1): 7-19.

Washademoak Environmentalists (2003) "Evaluation of the Canaan-Washademoak Ecosystem: Toward Conservation and Sustainable Management through Water Quality Monitoring and Riparian Zone Assessment", Unpublished report funded by NB Wildlife Trust Fund, and supported by Canaan-Washademoak Watershed Association and Environment Canada's Ecological Monitoring and Assessment Network.

Watt, W. D. (1973) "Aquatic ecology of the St. John River", Volume I: background, design, methods, nutrients, bacteria, phytoplankton and primary production. Report 15f for St. John River Basin Board, October 1973.

Zalewski, M. (2000) "Ecohydrology - the scientific background to use ecosystem properties as management tools toward sustainability of water resources", *Ecological Engineering* 16(1): 1-8.

Zalewski, M. (2002) "Ecohydrology - the use of ecological and hydrological processes for sustainable management of water resources", *Hydrological Sciences Journal-Journal Des Sciences Hydrologiques* 47(5): 823-832.

Section III:

Conclusions

International Comparative Studies on Water Resources Management in Diverse Ecosystems

Caroline King and Zafar Adeel, UNU-INWEH

The papers and case studies that are presented in this book offer a series of perspectives and insights on the challenges to be faced in the management of water and ecosystems to provide for human needs. The presentation of these studies and the issues that they raise animated lively discussions amongst participants at the international workshop where they were first presented. Over the course of the workshop, a group of water and ecosystem management experts from other regions took the opportunity to reflect on these presented papers, as well as on their own experiences, in order to develop a set of common statements concerning the challenges to be faced in water and ecosystem management at the global level. These findings are presented in this chapter as a series of four global observations on the existing challenges, followed by three essential elements to achieve the necessary capacity development to manage water and ecosystems and provide for human needs. A full list of participants in the international workshop is included at the end of this book.

Findings of the International Workshop Discussions, Hamilton, 14-16 June, 2006

1. Human well-being is impacted directly by water supply management and indirectly through effects of these management approaches on the broader range of ecosystem services.

There is a growing recognition that the provision of water supply for drinking and sanitation achieves reductions in poverty through improved health and productivity. In addition, the availability of water is essential to the achievement of many of the Millennium Development Goals, including the eradication of extreme poverty and hunger through improved food production. However, without careful management, increases in water extractions for human development can come at the expense of water needed to sustain ecological systems. The pressures placed on freshwater ecosystems can affect their ability to continue to provide services such as clean water supplies, waste assimilation, climate regulation, and the maintenance of fertile soils, vegetative cover and species diversity.

For example, in water abundant areas, water is easily polluted by discharges from human activities and livestock, and dams interrupt the flow of most large rivers, altering fish habitats and upsetting nutrient balances downstream. In dryland areas such as the Gulf countries, overgrazing has removed plant cover, impairing the ability of natural systems to circulate and retain water from rainfall. This leads to further water shortages, land degradation and loss of productive capacity.

The Millennium Ecosystem Assessment endeavored to assess the extent of the degradation of ecosystems around the globe. While the available data on the extent of the existing damage to ecosystem health was found to be incomplete, a clear observation was made regarding the disproportionate impact of the degradation of freshwater ecosystems on the lives of the poor, who are most often dependent on environmental services for their health, livelihood security and protection of their homes.

2. Better management approaches can be developed by using reliable information about the condition of and threats to freshwater ecosystems, as well as their impacts on human well-being.

At the watershed level, the interacting processes and balances amongst plant species, climate and water flows regulate the carrying capacity of freshwater ecosystems to withstand human induced alterations. In traditional societies, inherited management systems incorporate many generations of experience in protecting these balances. In response to new human demands for ecosystem services, scientists use experimentation and data on freshwater ecosystem processes to protect and extend the carrying capacities of ecosystems. Innovative approaches to management, such as ecohydrology, focus on understanding and manipulating the interplays between hydrology and biota. These techniques enable managers to control nutrient and pollutant loads in freshwaters and to regulate catchment hydrology. Such approaches can come at low costs, for example, allocating water in support of ecosystem functions, protecting water sources and conserving wetlands instead of installing water treatment plants, reforesting upstream areas instead of building new dams and flood barriers, recapturing nutrients from agrochemical runoff instead of applying more chemicals, and restoring the value of degraded land and water through phyto-remediation techniques and the rehabilitation of biodiversity.

Managers and scientists in protected areas and biosphere reserves can work with communities to apply such techniques to improve both the availability of economically valuable species and supplies of clean freshwater. Examples of the use of such management approaches, particularly for nutrient regulation, are presented in this volume from the Hengshui Lake in China, Lake Naivasha, Kenya and the Pilica River, Poland. Similar approaches may complement the development of basic infrastructure to provide water treatment and sanitation systems in developing countries, if applied and maintained at the appropriate scale by small communities. These approaches rely on the use of adaptive management techniques, community participation and the use of collected data on environmental processes.

3. Community participation in water and ecosystem management, through interventions that are in line with local priorities and support local livelihoods, is essential to the successful implementation of improved management approaches.

Many examples of successful community management at local scales were discussed by participants who attended the international workshop where the papers in this book were first presented. The experiences of community-level water management that they drew on during the discussion ranged from traditional water management practices in the Gulf, to the specialized knowledge of fishing communities in Asia. Examples of new developments community participation in data generation also demonstrated considerable potential. Volunteer data generated at the local scale through work with schools was observed to have already provided the basis for improved environmental management and policy development in the Waterloo region, Canada. NGOs and research groups can be effective in developing such activities with community groups, bringing scientific credibility, and a policy-oriented vocabulary to support the findings of community-based studies.

Experiences from the Australian outback to the prairies of Canada have demonstrated that an essential element is the development of trust between the community and researchers. Direct benefits to local livelihoods are widely seen as the key to adult community participation. An experience from the GAP project (Turkey) was shared, where a combination of incentives from different sectors of management, including subsidies and farm livelihood issues, encouraged farmer participation in community management meetings. Direct financial support for upstream community participation from downstream industries and dam users has enabled approaches to combat deforestation in various regions. Community-based approaches to data collection can create the basis for the development of such effective ecosystem management approaches, however, in order to develop a comprehensive knowledge of freshwater ecosystems, community volunteerism eventually needs to be replaced by a regulatory framework, a mandate to report, and a budget to maintain reporting.

4. It is essential to have adequate human and institutional capacity to successfully integrate the demands of water management, ecosystem conservation and poverty reduction. Such capacity is often lacking in most regions and the extent of capacity building required is also not well-known.
Political will is an essential element in the development of and maintenance of institutional capacity. In many freshwater systems, jurisdictional boundaries as well as sectoral interests can block integration between distinct agendas for socioeconomic development on the one hand (eg WEHAB), and water resources management across sectors on the other. In recent years, the conflicting priorities of the sectoral approaches to water supply and sanitation without consideration of ecosystem health has led to problems such as pollution by arsenic and fluoride, over-extraction of groundwater and salinization of water supplies.

In some cases, institutional development and integration of agendas is driven by extreme events, eg natural disasters that demonstrate the need for this approach. For example, in Ontario, floods on the Grand River led to the creation of a relatively powerful conservation agency. In other countries, such as that of South Africa (see Revenga in this volume), political regime changes and conflicts have led to the development of new integrated management structures and policies that focus on the management needs of freshwater ecosystems. In the absence of such extreme events, institutional change can be driven by the science community from the bottom up through the development and sharing of environmental data. For example, the success of the Mekong River Commission in improving management approaches can be attributed to its focus on information analysis and sharing between management communities. In South Africa, the need to develop, understand and analyze information has provided the force for new institutional development. On the Saint John River, Canada (see Munkittrick et al in this volume), persistent independent efforts by researchers led to the eventual remodeling of the Department of Environment sampling programme. Where scientists can influence the policy development process, such transformations can happen quickly, as in the case of the development of the Canadian Agricultural Policy Framework, which includes provision for early identification of environmental problems.

Essential capacity for water and ecosystem management includes human capacity within official management regimes, as well as universities, NGOs and community groups. In all of the case studies presented at the workshop, this capacity was considered to be underdeveloped to some degree, although some particular successes were noted in Lake Victoria, and the Saint John River (Canada).

Capacity development actions needed to improve water and ecosystem management

The following actions were identified by participants at the international workshop as priorities to improve water and ecosystem management:

1. Working with grassroots practitioners

Enabling point to point exchanges between researchers and knowledge-providers through research meetings and publications focusing on:

- identification of best practices in the use of environmental data collection for management approaches including ecohydrological or low-cost, community managed solutions
- development of structured analyses of applicability, including socio-economic benefits, valuation of ecosystem services and effects on policy
- dissemination of results to research and policy making communities

2. Institutional-level study of water and ecosystem management

Building on recommendations from case study presenters and in partnership with local research institutions:

- develop conclusions from workshop case studies regarding capacity development needs, including further development and testing of the hypothesis raised during the workshop that environmental data generation and sharing for scientific research and analysis can drive institutional development
- develop a research project including further coordinated desk studies and expert consultation to elaborate a typological overview of existing human capacity for management of freshwater ecosystems in selected regions, including critical identification of gaps, and the role of research institutions filling these by working with local communities and centres of knowledge
- collaborate with regionally-based experts to explore the wider development of findings through collaborative survey design and administration process

3. Developing an understanding of policy level linkages with science

Drawing conclusions from experiences and observations gained within the two previous elements:

- maintain a strategic overview of how to systematize the scaling up from the small-scale capacity building and research interventions to achieve institutional changes
- demonstrate to the international development community that the practical application of science and exchange of science-based knowledge can produce results.
- contribute conclusions to inform global assessments

The findings captured above were first presented in a report on the international workshop by UNU-INWEH. This report was circulated and discussed amongst the workshop participants and further refined until it was accepted by all as a common statement on the challenges of water and ecosystem management. These common findings are reflected in the papers that were developed for inclusion in this volume. While the workshop itself offered an opportunity for exchanges between researchers and knowledge-providers, the further development of the papers as they were prepared for this book encouraged authors to devote additional to the scrutiny representation of the institutional-level challenges, including capacity development needs.

Already, within this volume many policy-level linkages with science are highlighted in the papers and case studies. Carmen Revenga introduces this discussion with an example from the South African River Health Programme, instituted under the Water Act to inform policy decisions. The Canadian case from the Saint John River, describes a partnership between federal and provincial government agencies, private industry and NGOs centered around the research activities of the Canadian Rivers Institute. In this example, the linkage between science and policy has been forged from proactive work carried out and effectively communicated by scientists. A number of the case studies presented in this book address transboundary ecosystems: the Guardiana and Amazon Rivers, as well as Lake Victoria. In these cases, international policies for scientific collaboration are at the center of efforts to improve management, through joint research initiatives, such as those described in the relevant chapters of this volume.

As scientists and policy-makers in different regions of the world continue to work to build the capacity that is needed to address the challenges of water and ecosystem management, the challenges that they will face will be many and varied. The findings and experiences presented in this book offer some generic observations and recommendations on the key issues to be faced, and highlight successes already achieved. It is hoped that these insights may inspire both scientists and policy-makers to pursue efforts to provide water for human needs without compromising the ecosystem services on which these water resources will continue to depend.

Workshop Participants

Dr. Zafar Adeel
Assistant Director, UNU-INWEH
McMaster University
1280 Main Street West
Hamilton, Ontario L8S 4L8
CANADA
Tel: +1-905-525-9140, Ex. 24517
Fax: +1-905-529-4261
Email: adeelz@inweh.unu.edu

Prof. Luis E. Aragón
Núcleo de Altos Estudos Amazônicos
Universidade Federal do Pará
Campus Universitário do Guamá
66.075-900 Belém, Pará
Brasil
Tel: 00.55.91.3183-1231/3183-1951
Fax: 00.55.91.3183-1677
Email: aragon@amazon.com.br

Dr. Alex T. Bielak
Director, Science Liaison Branch/Directeur, Direction de la liaison scientifique
National Water Research Institute/Institut national de recherche sur les eaux
Environment Canada/Environnement Canada
867 Lakeshore Road, P.O. Box 5050
Burlington, (Ontario) L7R 4A6
Canada
Tel: +1 905 336-4503
Fax: +1 905 336-4420
Email: Alex.Bielak@ec.gc.ca

Dr. Sandra Brown
Institute for Resources and Environment
University of British Columbia
2206 East Mall
Vancouver, B.C. V6T 1Z3, Canada
Tel:+1- 604-822-9245
Email: sjbrown@interchange.ubc.ca

Dr. Patricia Chambers
Aquatic Ecosystem Impacts Research Branch
The National Water Research Institute
Environment Canada
867 Lakeshore Road,
P.O. Box 5050
Burlington, Canada, ON L7R 4A6
Tel. 905-336-4529
Fax: 905-336-4400
Email: Patricia.chambers@ec.gc.ca

Dr. Miguel Clüsener-Godt
Division of Ecological Sciences
Man and the Biosphere Programme (MAB)
South-South Co-operation Programme, (UNESCO-MAB)
1, rue Miollis
75732 PARIS Cedex 15 France
Tel: +33 1 4568 4146
Fax: +33 1 4568.5804
Email: m.clusener-godt@unesco.org

Dr. Ralph Daley
Director
United Nations University, International Network on Water, Environment and Health (UNU-INWEH)
McMaster University
1280 Main Street West
Hamilton, Ontario L8S 4L8, Canada
Tel: +1-905-525-9140, Ex. 24517
Fax: +1-905-529-4261
Email: daleyr@inweh.unu.edu

Prof. John J. Drake
Director
School of Geography & Geology
McMaster University
Burke Science Building, Rm 233B
1280 Main Street West
Hamilton, Ontario, Canada L8S 4K1
Tel: (905) 525-9140 ext. 24018
Email: drake@mcmaster.ca

Ms. Annette Dubreuil
UNU-INWEH
McMaster University
1280 Main Street West, Hamilton
Ontario L8S 4L8
Canada
Tel: +1-905-525-9140, Ex. 24517
Fax: +1-905-529-4261

Dr. Max Finlayson
Principal Researcher - Wetlands Ecology
IWMI
127 Sunil Mawathe, Pelawatte,
Battaramulla, Sri Lanka
Or PO Box 2075, Colombo, Sri Lanka
Tel +94-11-2787404 extent 1306
Mobile +94-777562217
Fax +94-11-2786854
www.iwmi.org
Email: m.finlayson@cgiar.org

Dr. Velma Grover
Project Officer
UNU-INWEH
McMaster University
1280 Main Street West
Hamilton, Ontario L8S 4L8, Canada
Tel: +1-905-525-9140, Ex. 24517
Fax: +1-905-529-4261
Email: grovervi@inweh.unu.edu

Ms. Caroline King
Project Officer
UNU-INWEH
McMaster University
1280 Main Street West
Hamilton
Ontario L8S 4L8
Canada
Tel: +1-905-525-9140, Ex. 24517
Fax: +1-905-529-4261
Email: cking@inweh.unu.edu

Ms. Leslie Kulperger
215 Spadina Ave.
Toronto, ON M5R 2T9
Tel: 416.929.6675
Email: lesliek@3web.com

Prof. N. F. Madulu
Institute of Resource Assessment
University of Dar es Salaam
P.O. Box 35097 Dar es Salaam
Tanzania.
Tel. +255 22 2410144 or +255 744 272335
Fax. + 255 22 2410393
Email. madulu@ira.udsm.ac.tz

Dr. Ian McCann
Irrigation/Water Management Specialist
University of Delaware, R&E Center
16684 County Seat Highway
Georgetown, DE 19947
USA
Tel: 302 8567303
Email: McCann@UDel.Edu

Dr. Colin Mayfield
Assistant Director
UNU-INWEH
McMaster University
1280 Main Street West
Hamilton
Ontario L8S 4L8, Canada
Tel: +1-905-525-9140, Ex. 24517
Fax: +1-905-529-4261
Email:
mayfield@sciborg.uwaterloo.ca

Dr. Kelly Munkittrick
Canada Research Chair in Ecosystem Health Assessment
University of New Brunswick
Department of Biology
Associate Director, Canadian Rivers Institute
220 KC Irving Hall, Tucker Park Road
PO Box 5050
University of New Brunswick
Saint John NB, E2L 4L5, Canada
Tel: +1-506-648-5825
Fax: +1- 506-648-5811
Email: krm@unbsj.ca

Ms. Stacey Noel
Research Associate
Stockholm Environment Institute at York
University of York
York YO10 5YW, UK

Ms. Anne Ollivier
Intern
UNU-INWEH
McMaster University
1280 Main Street West
Hamilton
Ontario L8S 4L8
Canada
Tel: +1-905-525-9140, Ex. 24517
Fax: +1-905-529-4261

Dr. Raul Pacheco Vega
CIATEC
Omega 201 Fraccionamiento Industrial Delta
Leon, Guanajuato 37480
Mexico
Tel: +52-477-7-10-00-11 ext. 1519
Email: rpacheco@ciatec.mx

Ms. Carmen Revenga
Senior Freshwater Scientist
Habitat Assessment Team
Global Priorities Group
The Nature Conservancy
Worldwide Office
4245 N. Fairfax Drive, Suite 100
Arlington, VA 22203-1606
Tel: +1-703.841.4513
Fax: +1-703.812-4975
Email: crevenga@tnc.org

Dr. Mark R. Servos
Scientific Director, Canadian Water Network
Canada Research Chair in Water Quality Protection
Department of Biology
University of Waterloo
200 University Avenue West
Waterloo, Ontario N2L3G1, Canada
Tel: 001 (519)-888-4567 x6034
Email: mservos@uwaterloo.ca

Dr. John Soussan
Research Leader
Stockholm Environment Institute
University of York
Heslington
York, YO10 5DD
UK
Tel: +44 (0) 1904 434745
Email: johnsoussan@hotmail.com; js47@york.ac.uk

Dr. I H Olcay Unver
Visiting Distinguished Professor
215 White Hall, Kent State University
Kent OH 44242, USA
Tel: 330-672-0563 / 330-672-0770
Fax: 330-672-2879
Email: ounver@kent.edu

Prof. Maciej Zalewski
Director International Centre for Ecology
Polish Academy of Sciences
90-364 Lodz ul. Tylna 3
Poland
Tel/Fax - (48 42) 665 5819
Email: mzal@biol.uni.lodz.pl

Dr. Guan Zhuojin
Beijing Hydraulic Research Institute
Haidian Chegongzhuang Xilu 21#
100044 Beijing
China
Tel: (0086)10 88416494
Fax: (0086) 10 88423808 / 0086-10-88416511
Email: guan-zhuojin@sohu.com